AF496479

We the People

Insight Editions

New York & Seattle

Also by Peter S. Temes

One School Now: Real Life at Lynn English High School

Great Books and Big Ideas in the Lives of Our Children

Against School Reform and In Praise of Great Teaching

The Just War: An American Reflection on the Morality of War in our Time

The Power of Purpose

The Future of the Jewish People in Five Photographs

The Teacher's Heart

Preface

For centuries, the height of complex technology was the *Katana* Samurai sword. The slightly curved *tamahagane* steel weapon has long been familiar as a symbol of traditional Japanese culture, built of layers of forged steel of varying carbon concentrations. Five hundred years before the invention of the electric light, tens of thousands of these extraordinary swords, delicately balanced works of art and craft, were used not only to kill but also to project a message about the sophistication and strength of Japanese civilization. Each sword represented centuries of cultural knowledge from metallurgy to human anatomy, directed by an individual. Personal training, discipline and adherence to the mystical honor code of the Samurai called *bushido*, "the path of the warrior," connected man, machine, and the culture from which the sword grew, century by century.

The blade cuts flesh – that's a big part of what it's built to do – but the Samurai's hand holds the blade skillfully. The handle, covered in leather, eel skin or shark skin, is tightly wrapped in a wound, narrow cord. A circular metal guard sits right above the handle's top, often beautifully crafted to depict gardens and cloud-dotted images of spiritual ascent. This is the *tsuba*, and the images above each chapter here are all photographs of *tsuba* from the collection of the Metropolitan Museum of Art in New York.

The *tsuba* is a perfect symbol of the intersection between the human hand and the evolved machine that can serve and represent us yet might also do us harm if we are careless. The sword's layers – forged red hot, the technique and tools handed down through generations – bring to mind the layers in the technology stack of progressively more refined and purpose-focused languages and applications that sit below functional digital systems of all kinds, ranging from a laptop's computer interface and kitchen refrigerator's digital temperature-management program to the systems propelling driverless trucks across the highway and monitoring the health of critically ill patients. These digital systems, the result of generations of craft and technique, accomplish extraordinary things for billions of us every day but will bite badly if we are not attentive to that critical spot where the human and the machine meet.

The *tsuba* has heft and most of all it has balance. Round or close to it, crafted either as a solid with a hole for the sword blade or more resembling a mosaic with open space interwoven in an intricate design, its weight is part of the sword's equation of force and a part of the Samurai's skill in moving that exquisite, simple machine through the air or through a standing adversary. Balance is almost an ethic embedded in the *tsuba*'s form and function, an implicit answer to the fundamental questions about how to live that follow all men and women through the ages.

*

As digital technologies help us do more of what we do, do it faster and with less investment of blood, sweat and toil, they inevitably turn us back to the questions we have been asking since the ancient Greeks: What is the good life? What do I owe my neighbor? What must we do in the face of injustice?

Ethics can surely be a useful concern for the perfectly isolated individual, but it blossoms in importance as we consider how to deal with others – what to give and what to take, what to do and what to avoid doing. A fundamental task of digital ethics is to balance the technological with the human. Certainly a joyous human task, at least for someone with a bit of faith in the future, is to help spur the conversations that bring different people from different places together to express and celebrate the beliefs that we can find in common: beliefs about the value of human life, about the value of community, and the penchant of the human species toward altruism. That is a great part of the work of digital ethics.

In a book I liked when I was a young man, a traveler on the road with his son tells the reader, "our main skill is not getting lost." That notion is important for this book as well: digital ethics is about not getting lost. A sense of where we are – where we start from – is key to this task. But more important is a sense of *who* we are, which I understand to mean mostly what we believe in. It's easy to be in a familiar place and feel lost if we don't understand the proper limits of what we might do in a moment or in a day. And it's easy to be lost if we don't know where we want to go.

The practical problems of coding ethics into our digital systems are large and important. This book will talk about those problems in some detail, and offer a few different frameworks and useful practices to address them. Just as important, it will call for a great conversation about who we are; what we value more, or less, when faced with inevitable choices; and where we want to go.

*

This book makes the argument that a civilization-wide conversation is essential in the age of digital systems and artificial intelligence. In our lives prior to the age of artificial intelligence, we often acted – and still do act – on values that we are not fully aware of, and we make decisions based on what feels right without reflecting much on where that feeling of rightness comes from. Today we must begin being more explicit about what we believe and why, so that we can code these values into our systems. We can't rely on the already inadequate notions of common sense or gut instinct when we live in the highly literal world of machines that execute code as their way of finding purpose and limits on their actions.

This book also reports on a substantial attempt launched by a number of leaders at a few of the world's largest technology, retail and healthcare organizations to put down on paper a set of ideas and beliefs that can help anchor the conversation that needs to spread across all of our enterprises, governments and associations.

The first part of this book tries to explain why the civilizational conversation about ethics is so vital; the second part tries to offer tools to help that conversation start and spread in practically useful ways.

Three documents make up this second half, originally drafted in early versions over a two-day "digital ethics summit" held in Seattle at the end of April of 2018, hosted by Northeastern University and attended by about 40 people working at range of companies and organizations whose names are widely known, as well as a few academics, entrepreneurs and not-for-profit organization leaders.

The first of these documents, a statement of aspirations, is meant to outline fundamental ideas that form the most basic layer of ethical belief. It is not a set of rules or expectations, but a statement of human purpose and appreciation for human life and community that talks about why we create technologies, how these technologies reflect generations of human endeavor and value creation, and at the most general level how the gains from technologies should enrich shared human experience.

The second of these documents, a statement of duties, expresses what different parties in the digital-ethics ecosystem ought to recognize as important obligations based on their roles as individual creators, companies, governments, and consumers.

The third document is a guide for designers and builders of digital systems as they navigate practical trade-offs in the course of their work.

These documents are not meant to be complete or final, and none is meant to be followed to the letter. Rather each of these three documents is meant to be an anchor, a common reference point for people working under very difference conditions at different times and in different places, reminding them of the purpose and potentials of digital systems, and always tugging them towards commonly held aspirations for the kind of world we want to live in as we move ahead in the evolution of this world at increasing speed.

These are not instructions as much as they are tests, reminding us to ask often and honestly whether the new tool, the new digital experience, the new line of code or code-making machine is doing what we want it to do in terms of those big questions that only become bigger in the arc of our progress: What is the good life? What do I owe my neighbor? What must I do in the face of injustice?

-   Peter S Temes

Foreword

I cannot think of a more important and timely topic in the world of business and technology than digital ethics – a topic that has, arguably, also become pivotal for the lives of billions of people.

For me however, collaborating with Peter on this book has also been a personal endeavor and a project which has allowed me to come full-circle. As an adolescent (and beyond), one of my favorite authors was (and still is) Isaac Asimov. I spent countless hours reading and re-reading his science fiction books and I remember being entertained and intrigued by his "Three Laws of Robotics" which were introduced in 1942, way ahead of their time and much before I was even born. They stated:

> **First Law:** A robot may not injure a human being or, through inaction, allow a human being to come to harm.
>
> **Second Law:** A robot must obey orders given it by human beings except where such orders would conflict with the First Law.
>
> **Third Law:** A robot must protect its own existence as long as such protection does not conflict with the First or Second Law.

The gradual introduction of a new Law Zero – "A robot may not injure humanity, or, through inaction, allow humanity to come to harm," was especially fascinating. How do you define "humanity" and how does one measure harm to humanity against possible harm to a human?

As I later started my university studies and elected to take some courses in neural networks, I remember asking myself whether I was wasting my time with something that was unlikely to have any practical significance. Fast forward a few decades and what was once science-fiction is uncannily close to reality. Technology advances are making AI very real indeed and something which directly impacts markets valued at hundreds of billions of dollars. AI is rapidly becoming a major geopolitical force.

At the same time, the democratization of access to and affordability of digital technology is bringing questions around digital ethics into the lives of billions of people. These questions are certainly challenges that I must consider every so often. I work for a company at which a core value is to "change things for the better," and where for the last couple of years, we've tried to crystalize and agree upon some practical guidelines for the designers, data scientists and software engineers who work on hundreds of projects around the world, especially around data.

Not too long ago, "the bigger the data, the better" was the norm in our industry. There was an almost childishly naïve drive to add ever more data sources, to combine and "enrich" your big data in the hope of mining unrivalled competitive insights. But what if these insights could limit an individual's opportunities and possibilities? Shouldn't the highest priority be to respect the individuals that this data describes and affects? And to add more context, How was this data collected and why? Who is responsible for understanding and affirming the provenance of this data?

What if this data is not only used within the context and intention in which it was collected, but rather re-purposed?

What if data is assumed to be "neutral" when in actual fact it carries a history of human decision making, which will shape the consequence of its use?

These are no longer questions which just a handful of born-in-the-cloud organizations need to engage. Every organization with a modest IT budget (or even just an individual with a credit card) can today buy computing power (and often data) that has the potential to impact lives.

Who will be invited to a job interview?

Who will have access to transportation?

Who'll have a lower risk of accidents in a mine?

Who will get the best prices and the best service when they shop – for a T-shirt, or for health insurance?

One can, and should, work on creating the practical, actionable "coding guidelines" that technologists and business executives need to do the right things and to change things for the better, but these are not enough.

Working together with Peter on this book has helped me deepen my commitment to clarifying and sharing fundamental values and principles to keep our technology-driven society in balance. To use an American civics analogy, the Constitution – the operating manual for how to run the American government – needs the Declaration of Independence, containing the aspirations and values that our government exists to serve, and vice-versa.

This book is just the first step, with much more work required. We hope you will find this book informative and inspiring. Come join us and the community as we take the next steps!

-    Florin Rotar

Contents

PART ONE

PART TWO

Introduction

The Civilization-wide Conversation, Beginning with a
Man and a Dog

Imagine the factory of the future. The scholar and writer
Warren Bennis sat down some years ago to do exactly that. He
thought hard and decided that in this factory there will be only
two employees – a man and a dog. The man's job will be to feed
the dog. The dog's job will be to keep the man from interfering
with the equipment.

This little thought exercise captures a lot about the challenges
of digital ethics today. Computers and the automated systems
they run have already helped us live longer, healthier lives,
communicate instantly across the world, and live better by

almost any material measure. Digital systems are faster, smarter, and more capable than ever – by orders of magnitude compared to just a few years ago. The rising ability of these systems to learn from astoundingly large amounts of information means they're getting much more powerful at a much faster pace. How long we live, how well we live, where and with what pleasures and creative power – all this is changing, and the change is rich in positive potential.

The man and the dog in Bennis' story suggest on the surface that people will become obsolete as workers – or at least as factory workers – in the age of digital automation. And as driverless trucks make truck drivers obsolete, as better digital systems reduce the need for everything from short-order cooks to bank tellers to accountants and even to software engineers, we'll be left feeding the dogs who keep us in line. A bleak scenario indeed. But there is more in the factory-of-the-future story below the surface.

The man is there to feed the dog, yes, but most important is that the man *is there*.

His presence is an important physical, social and spiritual fact. Human presence is necessary. Just as the dog watches the man to keep him in line, so the man watches the machines – and watches the dog. The watcher watches the one watching him. Many of us have seen this as people we know and care about become more prosperous: a story of progress we can measure year by year in some cases, and generation by generation in others. The working man or woman who transits from earning barely enough to having *more* than enough changes not only materially, but in terms of the story of his or her own life. Work becomes something different as people earn more money and worry less about where they'll eat and sleep. The ways that we work, the shelter we find and cultivate, the food we eat – the meaning of all of it is up for grabs as the story of our lives becomes a different kind of story. "Can we survive?" becomes "How should we live?" And the making of things – growing food, building shelter, crafting tools – becomes more than a series of means to reach obviously necessary ends. The man in the factory is not only feeding the dog, he's connecting the creating of goods in a totally autonomous place, by machines that work on their own, to the human story. He is a witness, a connecting point between our systems and our lives; between our systems and our stories about ourselves, about what matters most, about the purpose of things, the purpose of our lives, and what is right and what is wrong.

The story of the man and the dog in the factory of the future suggests that the machines we make can and should never be free of us. We are drawn to structure, watch and witness what they do, and inevitably to bargain with the dog watching us and depending on us for its food, to allow the unavoidable intervention of the human hand when we see the machines go astray. The presence of the human is the link to human purpose – whether to build cars or make sweaters or accelerate the power of semiconductors – for which the machines were built.

This man and this dog are important to keep in mind as we shape the practice of digital ethics. Human presence brings with it not only the link to human purpose but also the opportunity for the system to be adjusted by the human hand. That, in the end, is a good thing specifically because the efficiency of systems must never be an end in itself. Efficiency *doing what* is the question we can never let go of even as the humming logic of machines and digital devices encourages us to loosen that grip. The dog's presence in fact hints at the inevitable have-it-both-ways challenge of keeping the human hand close enough to stop the machines but highly incentivized not to. That is the tension that digital ethics as a practice must balance – the tension between the human purpose of the non-human tools and systems, and the dark potential of systems and machines given poorly framed tasks and not given proper limits.

Getting beyond the gut feeling

Actual humans make most of our ethical decisions quickly based on gut feelings that reflect values we often don't even know we have. Yet now we must know, so we can tell digital systems how to make these tough choices.

Should we value all human life equally, for example, or prioritize the lives of people who have entrusted us with their safety?

When is it better to help a person in need in the short term rather than protecting the natural environment in the long term – in cases like increasing crop yields in areas with food shortages, or lowering prices for food or fuel when people have great need in a crisis?

When digital systems make decisions like this almost instantly and without human intervention, what principles about right and wrong will they draw on? What deep-learning algorithms are constantly evolving? What values and what biases are being reinforced?

We need to begin talking about ethics a lot more, across our civilization, right now, to make sure that we "code in" what we really believe about these tough choices.

The ethics layer of the technology stack

To do this coding-in, we need to keep in mind that digital systems are built on a "stack," layers of technology going from the more general to the more specific. We must make sure that the first layer in every technology stack is an ethics layer. Thinking and talking about ethics – a lot – can and must become part of the software engineer's, data scientist's and system architect's tool-kit.

From instance-based ethics to principle-based ethics

To make this shift in thinking and acting as we build digital systems, we should keep in mind the critical shift from instance-based ethics to principle-based or rule-based ethics. Instance-based ethics is what we practice when we see a thing or an action and say, "that's wrong." Let's say we witness a big kid in a schoolyard come up on a smaller kid and grab his ice cream cone. The big kid exults, "Now that's mine! I'm gonna love it! It's my third one today!" We see this and say, "That's wrong," secure in our sense that we've seen something out of bounds. People often explain this kind of ethics by saying "It's obviously wrong," or "You just have to look at that and you know it's wrong."

The high-point in instance-based ethics came in a 1964 U.S. Supreme Court decision about pornography. Justice Potter Stewart, in his concurrence with the court's decision to exclude a specific bit of film from a legal ban against "hard core pornography," wrote that "I know it when I see it, and the motion picture involved in this case is not that." Many of us would agree with Justice Stewart. The film in question, director Louis Malle's *The Lovers,* is considered a serious work of art, sexy but not pornography by most measures and far from hard core in either case. Se we can look back more than fifty years and feel that Justice Stewart got it right. But he also made the profound limits of instance-based ethics clear.

Can we apply the underlying standards that lead us to good ethical judgement in *this* case to other cases? Short of conjuring the ghost of Potter Stewart, how are we to judge the next film? And just as importantly, the case of *The Lovers* was an extreme example, not hard to categorize. For a closer call, we'll need to reflect more carefully on what exactly makes a film fit in the objectionable category or not.

The Potter Stewart standard – "I know it when I see it" – keeps the principles that underlie ethical judgment inside the judge's head. It's a practice that makes the judge more important and more mysterious. We can't double-check his application of agreed-upon principles because he gives us only the judgement, and no glimpse of the process through which he arrives at it, in some ways concentrating the power he executes by keeping its methods and driving values private while its end result affects us all.

This is less a democratic exercise of state power than a judgement based on stated principles would be. It's also *slow*, expensive and opaque. Imagine the need to judge a hundred films, or a thousand, in a day. There would not be enough of Potter Stewart to go around, if our aim is consistency in restrictions on works or art or culture that might be pornographic. If we endowed the process with a hundred or a thousand Potter Stewarts, then we might move quickly — but if each "knows it when he sees it," we'll have a hundred or a thousand different standards enforced. If, instead, we can clarify what values Potter Stewart was actually applying to reach his judgment, the process can be faster, cheaper, and less dependent on one man's mood. We can move toward algorithms that can be trained to apply these principles, while fundamental questions about how the data we use in our "hard-core" digital tools is collected and why, and the evolving purpose of the film-evaluation program, remain vital human concerns, reviewed and discussed regularly. Humans do the values-laden thinking work; systems apply those principles at scale.

Instance-based ethics leads to conversations like this:

"What did you think about Bob's decision to fire Fred?"

"I thought it was wrong."

"Really? Why?"

"Well, just look at it — it's so clearly terrible. Just wrong. Bad. Anyone could see that."

"OK, but if we want to protest this, what should we say?"

"Say it's wrong because it's wrong – we know it."

"How do we know it? I mean, I agree – but we can't just protest because it's wrong."

"Are you kidding? That's the only reason to protest – because it's wrong. It's so wrong, totally wrong."

"Sure, but if we could say that Bob fired Fred for a reason, that there was a reason it's so wrong, like it was because Fred has different political beliefs than Bob, or Fred found out that Bob was stealing money so he was fired to keep him quiet, or that there's some prejudice here about where Fred's from or how he looks or something like that, we can get the big bosses to see that Bob's wrong, and maybe they can bring Fred back."

"Sorry – this is just wrong. I know it's wrong. I know it when I see it. That should be enough for anyone. Anyone can see it."

Instance-based ethics in a machine world makes little change; it mostly amplifies the challenges we face. Principle-based ethics makes more change, with more accountability, and forces the hard conversations to clarify and negotiate values when they clash. In the long run, principle-based ethics does far more good.

Deciding on the spot

Instance-based ethics requires a person or a group to decide on the spot whether something is right or wrong and keeps hidden the underlying principles that guide us in making these decisions. Rule-based or principle-based ethics give us the chance to openly apply principles in order to reach a decision.

This helps us make more consistent decisions across large numbers of deciding-points. And while the "I know it when I see it" school of ethics might seem to be a fast way to make decisions – just take a look at what you'll be judging, and then you'll know the answer – it risks amplifying bad judgment and unconscious biases as it scales. Either the judgments of individuals will not fit with the values and expectations of many others, or large numbers of people will have to hand-rate the issues and incidents in question to ensure that the judgements reflect the gut feeling of most of the group on whose behalf these decisions are being rendered. When the rules of what meets the "good/yes" standard or the "bad/no" standard are not stated, a degree of chaos is more likely than consensus (as the case of pornography in the U.S. in the decades following the Potter Stewart decision illustrate).

With rule-based ethics, once we've done the hard work of agreeing on what the principles *are*, in many cases we can automate the process of applying them, even the point of allowing our digital systems to be deciders of good/bad or yes/no. In fact, these systems are already making billions of decisions like this each day across the world – but unfortunately in many cases, without careful consideration of whether the rules being applied (or in too many cases, *not* being applied) are truly what human communities believe.

When principles need changing, best if they're visible

Automated traffic-signal systems, as one example, are turning lights from red to green based on timing rules that once made

sense but no longer do. If the rules that govern the turning of
the stoplights from red to green are clearly visible, we can see
whether they continue to make sense as the world around them
changes. If the rules are hidden, it's not only harder to change
them, it's harder to notice that they need changing.

Stopped on an entirely deserted road in the middle of
Brooklyn's thickly wooded Prospect Park after midnight on a
Tuesday some years ago, Peter sat behind the wheel wondering
whether he should just blow through the long red light that
made him feel like a target in a high-crime area. He was held
suspended in that bit of utterly empty road because the logic of
the light's red/green timing was based on a model of traffic flow
obviously off base. In later conversation with a city employee
tasked with monitoring the operations of the vast network of
Brooklyn's traffic lights, he learned the operating philosophy of
the system's human minders: "We know when a light needs to
be re-timed because we get complaints. When people are
complaining, it means something's wrong and we check it out.
When no one complains, nothing's wrong."

That was instance-based ethics expressed in a near-perfect
form. Because the principles and rules aligned with the human
purpose of traffic lights were hidden even from the people
running the system, drivers with the loudest political voice were
assisted (though in non-systematic ways) while others were
ignored – those struggling with the broken system who were
small in number (post-midnight park drivers) or without the
habit of complaining ("I'm from the Midwest; I was trained not
to complain") or with some good reason not to trust municipal

officials. These people are harmed rather than helped by the power of the automation that blinks the red light to green and the green to red in its baffling rhythm.

The technology behind the traffic lights in Brooklyn is decades old. It's slow moving. It can and does have life-and-death impact, but in slow motion compared to the emerging world of super-automation we're living in today. The harm done by unstated or impossible-to-comprehend rules shaping the behavior and impact of systems is being amplified by more powerful automated systems and more powerful artificial-intelligence engines. And this harm is likely to become vastly greater as the years unfold unless we can fill the holes so clearly present in that red light in the middle of the empty road through a park in Brooklyn. We need our systems to contain, to be limited by, and to make visible principles that people can believe in because they're based on a shared sense of values and purpose.

We need to make these principles clear, and to enumerate the conditions under which exceptions are allowed, and we need to do this at scale, starting right now. The communications piece of this equation – the making clear of what the principles and rules in fact are, and what they exist to accomplish – is particularly important.

Playwright David Mamet – the author of *American Buffalo, Glengarry Glenn Ross, Wag the Dog* and many other, often dark, works focused on the hidden schemes that organize ordinary experience and dictate much of what seems to be the natural order of things – talked in his book of essays *Make-Believe Town* about the particular kind of fear and diminishment that comes from an a rule-driven society or institution that does not tell people what those rules are until, one by one, they are broken. It's a brutal kind of aversion therapy – averting people from trust, from creativity, from exploration. Too many of our digital systems operate with similar effect. That late-night red light had *some* logic to it; or perhaps not. Break the evident rule of stopping at the red to discover whether it might be enforced in that place at that moment. Or just wait until eventually – hopefully? – it turns to green. What gets lost is more than time; trust in systems that hide their logic is hard to maintain other than through fear of what might happen if we cross them, when getting the principles that drive them aligned to our beliefs and sense of purpose – and communicating the rules that follow from those principles – can do the opposite, and reinforce a sense of technology, culture and public space all working toward some kind of common good that each individual can participate in.

A too-quiet conversation

We're already in the middle of the civilizational conversation we need – but we do it too quietly, too sparingly, and too far from our daily work and play. In many ways, we are already (in fact, *always*) in the middle of the process of agreeing on and expressing fundamental human values and shared human purpose. From the most ancient tribal codes to the writings of Greek, Roman and early African cultures to the many millions who address, today, questions of right and wrong, best possible outcomes, and what we consider the good life to be – in classrooms, in churches, temples and mosques, around family dinner tables – this is work we have been doing in our communities for as long as the human race has existed. Civilizations have risen and fallen based on the complex interactions of different value systems. Making war and making peace, curing disease, building cities and peaceful commerce all reflect the success or failure of shaping, communicating, and reshaping community beliefs and ethical rules.

Knitting together the understanding and compromise that allow different families, tribes, nations and civilizations to communicate, to share resources and common public space, and to build human community at a large scale is the fundamental task of human progress. Today, the shape and the scale of this task are strikingly different and, in some ways, more urgent than they have been in the past. Digital systems connect different communities more deeply and more intimately than ever before and have the potential not only to increase the material riches of the world but to make those riches more available to more people than ever before.

How can we bring these conversations and interactions that shape and reshape our common values up to the level of visibility and speed that can match the increasing pace of technology's advancement? We must begin with the most fundamental values.

Start with values

Values are answers to questions – not arguments that build off starting points of beliefs, but the bedrock beliefs themselves. More than consensus or points of agreement, values are the compass points, fixed positions that we test our actions against, not always getting those actions into total accord with our values but pulling and tugging those actions toward the values we hold.

To match a meaningful set of values to a fast-moving, complex society open to the world and built of many elements from far-flung sources that come together to form a complex but connected whole – *e pluribus unum* – the values at the core must be few but firm. A long list of rules is not a foundational set of values, nor is an easy avoidance of the hard work of forging common belief. *History is written by the winners* will not do as a way-finder. Actual answers to a small set of core questions are necessary to build upon. These answers must be, on their face, hard to reject outright, and should prompt further thought and nuance when applied to practical actions in the changing world.

These three questions are clear and fundamental. Answering them is a good exercise in the process of thinking through the principles that guide our actions and that should guide our digital systems. It's hard to answer questions like these not just for ourselves but as cultural norms that our systems should carry forward, but that's the fundamental digital-ethics task. Just doing this thinking work makes it clear, as well, that an important part of digital ethics is to answer these questions, and then to make sure that we give each answer some important limits or footnotes. We'll need to be quite crisp on whether a principle is one that needs to be always relied on, sometimes and under some circumstances, or in other ways limited. More about these shadings later; doing the thinking work to start answering these questions is the starting point.

The first question is about the value of human life; the second is about privacy; and the third is about equity – the sharing of the gains created by shared knowledge and shared work.

1)   **The Value of Human Life.** The first issue we have to clarify is *how valuable human life might be – and which human lives might be more valuable than others.* This most-fundamental question informs so much of what we ask digital systems to do.

Consider:

What do we tell the driverless car when we engineer the algorithm governing its behavior – protect passengers first, or value their well-being no more than anyone else's? Faced with the classic "I'm going to crash into *someone* – but I can swerve this way or that" dilemma, do we program our driverless vehicles to value one life over another? A child's life over an adult's life? A parent over a non-parent? A scientist over a barista? A nun over a wealthy man? Or should we just count up total lives in each alternative and decide based on numbers?

At what point do we say that helping a human – or a community of humans – live is more important than reducing environmental harm, or harm to animals or harm to a clean-water ecosystem? Do we need to put a thumb on the scale in favor of human life rather than an Earth-first agenda?

2)   **Privacy**. As digital systems become vast repositories of personal data – data even we as individuals don't know about

ourselves – *who gets to know what about whom* will become even more of a hot-button issue.

How our images, our creations (stored online more and more), our names, and records of our actions, our health and our finances are used, and by whom, will demand much more thought, discussion, agreement and coding-in to our digital systems than they receive even today.

How and when our devices guide us, warn us, and intrude in our lives will become critical. We need to clarify the values we hold that will set limits on these devices, empower individuals to control and benefit from their personal data, and write these values into the first layer of the stack of every digital system we create or update.

3)   **Equity**. As digital systems help us create more material wealth, who will benefit from that wealth, and how equally should those benefits be spread?

Bring a robust online learning tool into a classroom of a dozen students or more and you can see the effect: the strong get stronger. Students already succeeding are likely to apply the skills and motivation they've got to tools that bring – literally – knowledge from across the globe to their desktop, and become even more successful by many measures. Students who aren't doing their reading won't benefit when what they *might* choose to read expands a thousand-fold. While this might look like the rise of a digital meritocracy, in many ways it extends the advantages and reinforces the biases and disadvantages individuals are born into.

In cities like San Francisco, Seattle and Denver (or Stockholm, Sydney or Singapore), the same dynamic plays out in different terms: Who benefits from living in a richer city? While more wealth is clearly washing through these towns, the split between the top dogs and the rest creates social fissures that clearly do harm.

The question of equity asks us, as well, to think about the future of work. As less human work is needed to produce the goods and services we want, shall we spread the newly available leisure time evenly across society? Shall we see those who make and most skillfully use digital technologies benefit more than those who can't, won't, or simply don't?

With every digital advance, how much should we be thinking and talking about equity and fairness? How much should we let the chips fall as they may?

Three proposed answers to start the values conversation
These questions are vital, and the answers offered here are a starting point in the conversation that leads to values we can collectively embrace and encode on our systems.

Perfect agreement is not the goal. Enough agreement on enough of these critical issues to enable us to escape the default of *no limits* on our systems, or *arbitrary* limits, or the limits that suit mostly the code-writers or their investors or salary-payers, is the goal.

Proposition One: All human life *is* equally precious, but when the hardest choices are made, the lives of children should be valued over the lives of adults.

Proposition Two: Privacy matters more than most organizations in North America recognize today. Building many more points of explanation and permission-seeking for sharing of personal information is important. Anyone whose data is used to create value to others should be seen as a high-priority beneficiary of that data – the folks looking at and analyzing your data *owe you something* if that data creates value for others.

Proposition Three: Equity should be a high value, but not the highest. The well-being of all should be more important than equity – so that we don't "level down," or attain less well-being for everyone in order to avoid some having too much gain than others. Inequity can be a driver of ambition and incentives for creation that serves all. Yet extreme inequity is inherently harmful to the group, and in many cases materially unconscionable for the bottom dogs. Balance is vital; measurement and awareness of equity impacts might be a good element to add to every digital system.

Crafting palatable common positions

Each of these propositions deserves a great deal of discussion, debate and revision. The itchy feeling many will have about a position they don't quite agree with has to be balanced by the itchy feelings others get based on similar, opposing or simply different perspectives.

The crafting of some palatable common position is absolutely essential on these fundamental propositions. We clearly need to step forward and frame our digital systems within limits that reflect our values about human life, privacy and equity. We do currently address these vital principals in fits and starts, here and there, without clear agreement across the many perspectives among creators, users, and let us say *citizens* of digital systems – those effected daily and deeply by them, even if unknowingly.

The consent of the governed

The *Declaration of Independence*, published in 1776, includes this striking statement: governments derive "their just powers from the consent of the governed." This statement speaks to a great many civic and legal debates today. The governed are not only a nation's citizens; not only the men; not only people of a specific race; not only people with proper papers. The governed are everyone within the sphere of a government's control. Treating its citizens well, but non-citizens poorly or indifferently, will not make a nation just.

Put another way, the governed are those who bestow moral status on the government. Similarly, all those effected by digital systems are the measure of whether we are getting things right at the ethical layer of our technology. Not merely those who create or own or vote or are sought out for their opinions – *all* who are touched by digital systems are the measure of decency in this case.

The three documents that begin the second half of this book are attempts to draw in as many of these people, these digital "governed," to the conversation about digital ethics. These documents seek to articulate basic principles upon which we can build, and test, the ethical rules our systems will follow; they seek to address specific roles of different people based on how they engage digital systems through social or technological work and experience; and they seek to guide coders and architects of digital systems to make the building-in of values and an ethics layer in all systems a matter of habit and best practice.

Who should discuss these documents, help them evolve, embrace and endorse them? Everyone. Every company, every creator, every user every person touched by these systems. If the pages to follow in the first part of this book do what we hope they will do, the importance of this work should become more and more clear and the reader should have the very specific feeling that *it's time to do something about digital ethics*. The doing involves hard work in talking about ideas, ideals and values, beginning right now.

Chapter One

How to Make the World Better with Technology,
Without Technology Running Away with the World

The end of the human race?

A large number of well-informed people warn that at some
point smarter and smarter systems will start making their own
decisions about their purpose – about what they should be doing
for us and to us – and then acting on them without input from
actual people, whose ability to process information and make
decisions is just too slow for the automated systems to be
waiting around for when they can simply act and get things done
instead. No less a source than Stephen Hawking, not long before

his death in 2018, wrote that "The development of full artificial intelligence could spell the end of the human race. It would take off on its own and re-design itself at an ever-increasing rate. Humans, who are limited by slow biological evolution, couldn't compete, and would be superseded."

This is a fascinating proposition and it's worth paying a lot of attention to. But this dark possibility only happens if we choose to let it happen.

Every computer system we create is built in layers, all adding together to a full stack. Each layer includes not only instructions for how systems can and should get things done, but clear limits about what not to do. Only when we fail to set ethical boundaries based on our moral values can these systems fill that empty space. The problem is that far too often, we do indeed fail to write ethical boundaries into our computer code and AI algorithms. Now is the time to stop that failure.

Stopping is at once complex, and simple. The complex part comes in two flavors: the complexity of making sure that code accurately and consistently includes ethical limits, and the complexity of people agreeing upon and being aware of their fundamental values in the first place.

This second point – our need to be much more aware of what, exactly, our values are – is the key to what should be a great emerging age of ethical thought and exchange as a fundamental part of our cultural conversation. That is the promise of digital ethics, and we need to begin keeping it.

Systems already make better and faster decisions than humans – when the rules of the game are clear

The massive amounts of data flowing through computer systems that do things like monitor water quality, operate driverless vehicles, or even set prices for hotel rooms are already too big for humans to grasp and use for making decisions as fast as computers can. But at some point all this data does more than help computers decide whether to kill more bacteria in the drinking water, slow down quickly on the highway, or cut the price on a room at the Hilton. At some point the computer knows so much more than the programmers who set it in motion that – if we let it – it can scope-creep and assign itself new tasks.

Consider this not-too-distant possibility:

A massive amount of data about the water in a reservoir alerts a water-quality system that too much auto traffic on a road nearby is leaving too much motor oil on the road surface, which in turn is fouling the water. Given weeks or months, a human water-quality inspector would likely notice too, and write a report for her boss, leading to a meeting and some weighing of options about what to do – all moving at the speed of human institutions and local government decision-making.

But let's assume that the road is controlled by a computer program, and let's assume as well that this simple program is part of a "smart city" collection of systems that share data with each other even when humans are not watching. You can see this in some areas already where automated systems might, for example, close roads not only because of ice build-up but also because they know how many or how few cars are leaving the parking garages nearby or heading closer from feeder roads. Within a few minutes of the water quality crossing the threshold of unsafe levels of oil residue, the water-quality system might reach out to the road-control system and close the road, based on the "highest good" for the water-quality system being the preservation of clean water.

A big potential problem here is that one system's highest good or supreme mission might ignore its complex effects on other systems it touches. Focusing only on water quality makes sense when one computer system can order up new cleaning agents to be put in the water – but when that system can also shut down roads that lead to hospitals, as an example, we need to make sure that the newly powerful impact of systems working together to shape whole ecosystems of the built and natural environment is balanced by a newly powerful set of ethical trade-offs and limits in these systems.

The Ethics Architecture

There's a system-architecture issue here we should not avoid, though we often do. We might want to call it an "ethics-architecture" issue: when several different systems each with a specific function and purpose link up together, we need to look carefully at when and under what conditions the purpose of one system becomes more important than the purpose of others, so that the others can stand down and let the higher-purposed system do its work without interference.

The highway signaling system, for example, might be allowed to override the water quality system's highest good when an ambulance stands ready to take a very sick child with a highly treatable illness to the hospital. From an instance-based ethical perspective, this is an easy call. The good of the child is appealing in a direct and visceral way. Children are precious; we generally love and feel duty toward them more than we do toward lakes. But the principles at issue are complex, and that complexity is important to recognize. Clean water is vital for *many* children's health, so maybe we should slow down the leap to prioritize the ambulance's access to the road. But then we can see that in this instance a great deal of good accrues to the child, and only a tiny amount of harm accrues to the lake.

Start with purpose, then delineate exceptions

We can imagine exactly the kind of logic that goes into writing a simple computer program. We start with what we want to get done. We then tell the computer what to do in a series of small steps, but we also map out alternatives: if THIS happens, then go and do THAT. But if THIS OTHER THING happens instead, then don't do THAT, but do this THIRD THING.

It's certainly true that today, algorithms are trained to make decisions like this that are far more nuanced, based on pattern analysis and data classification, and often with more opaque logic. Still, we must ensure that an ethics layer is firmly present even – especially – in this new algorithmic world.

In the case of the road near the reservoir, we start with a purpose for the system – protecting the quality of the water. But we recognize that this purpose is part of something larger: preserving an ecosystem that helps keep the natural environment healthy and supportive of human life. So we rule out an uncompromised purpose: we don't want the water-protection system to protect water quality at any cost, because that cost could potentially do harm to the even greater purpose that hovers above water protection. Keeping the water in this reservoir clean at the expense of destroying thousands of acres of old-growth forest and cutting off access to hospitals would be good for the narrow purpose but bad for the higher purpose of healthy *water as part of a healthy environment supportive of human life.*

Now we have to code in a number of exceptions: prioritize the health of the water up to the point that we begin harm the larger ecosystem to a meaningful degree; prioritize the health of the water up to the point that we might harm a person or people to a meaningful degree. Of course, we'll have to define "meaningful degree" – a child's preventable death is certainly a meaningful degree of harm. A missed day of fishing on the reservoir might not qualify, but might add a bit of weight on the scale.

The reason to prefer a principle- or rule-based ethics over an instance-based ethics in this case should be clear: the catalogue of examples of harm to the larger ecosystem and human experience would have to be almost limitless. Making the right decisions will require us to express a set of rules of principles that the system can apply to any circumstances – including circumstances impossible to imagine today but which might well arise tomorrow. This not only saves enormous amounts of time but also makes a more sustainable bit of system software: the world we might describe instance by instance will change beyond our expectations over time. Yet the essence of right and wrong, of the relative importance of fundamental values and virtues should not.

Another example: years ago as an aspiring writer, Peter was fortunate enough publish a few small articles in the *New York Times*. He was thrilled. In part because he made his living mostly by teaching and consulting, the small amount of money paid for these articles did not mean much to him. The exposure and the chance to be recognized as a writer for the most widely read serious newspaper in the U.S. was more than enough reward.

Years later, he received a note from the *Times* asking him to sign a document that gave the newspaper the right to put his articles from those days up online. In addition to permission to post the articles, he was asked to sign away any claims for further compensation. He would not be paid any further for the use of his writing – long completed and paid for – in this new medium.

Over the years on a few occasions he'd been mailed a small check – seldom worth more than the cost of a decent meal – from either the *Times* itself or a publisher the paper had made a deal with to reprint one of those long-ago articles. In every case, he was delighted: his work would find a larger audience, he could by his sweetheart a nice lunch, and was continuing to benefit from by association from the publication that many referred to as "the newspaper of record."

The precedent was certainly there for the paper to offer a few dollars – let's say the cost of a couple of sodas in Peter's case, instead of the full-on subvention of a dinner for two – in order to add the rights to publish his work online to the already-paid-for rights to print his work on newsprint and put it in libraries on microfilm. Still, the *Times* faced thousands of living writers whose work had appeared at least once in the paper. Negotiating new use rights for old work with those writers would not only ring up quite a tab but also create a monstrous clerical task of tracking and payment-processing. The newspaper instead gave a choice to all: give us permission to use your work free-of-charge online or you will disappear from the online archives as though you'd never written for the paper at all.

For a low-level player on the stage of national news and opinion like Peter the choice was easy: take the rights to the work, please, and keep me in the club. In fact, send me a bill and I'll buy *you* lunch to make sure I'm included. For other writers, especially those full-time scribes who might have published hundreds of articles in the paper and relied on every bit of earned compensation to pay their rent and feed their families, the choice looked quite different.

From the newspaper's perspective, the dilemma did not seem like much of a dilemma at all. Most of the agreements writers had entered into over the decades in question offered a writer's fee in exchange for the right to print the article in a current edition of the paper, to keep it in the archives, and to reprint it if the occasion came up. In some cases, contract terms were different for high-profile writers. And certainly, over the decades, the wording and expectations changed a bit now and then. Yet the newspaper would have been within reason to assert that it had no way to imagine that something like the internet would emerge. When the *New York Times* was buying the right to publish an article, it took for granted that it would have the right to publish that article in future formats even if it could not specify what those formats might look like. That's why publishing agreements today often include clumsy-sounding but hopefully comprehensive language about the publisher paying for the rights to publish in "future mediums of any kind, including those not yet imagined." Thus the awkwardness of instance-based ethics: feeling the obligation to anticipate any and every case that might arise, we are backed into the corner of making specific claims for the things we can't yet imagine.

Far better, of course, for these agreements to anchor in principle than in the impossibility of exhaustive examples. If the principle here is that the newspaper pays once for the right to all future value to come from the writer's work first printed in the paper, that creates a kind of algorithm that we can apply in future cases with relative ease. If instead the principle is that the writer is being paid for use of the article today and tomorrow, but not forever, that's a different and opposed principle but one that also brings a good bit of clarity to the case.

If we can operate at the level of principle-based ethics instead of instance-based ethics, we'll spare ourselves the tedious work of deciding how to apply agreements in the past about instances that don't fit the changing world. We might have more honest and efficient work applying principles – like "writers should share, over time, in the money their work generates over time."

Purpose and trade-offs: some questions we must answer now

Important principles we need to code in to our systems can be revealed by imagining instances that make sense today; by tackling them based not on coding in proper machine behavior in the specific instance, but based on a larger principle that can be applied in novel situations, we leap from the limits of instance-based ethics to the future-friendly sphere of principle-based ethics.

Consider:

When a "smart" building security system has the option to unlock a secure door on a dark night after a microphone picks up banging on a door and a plea for help, what does it need to know in order to act? And *why*? (It's the *why* that leads us to the principle).

When interlocking systems at the heart of newly emerging "smart cities" balance human freedom with environmental impact, or human access to the things that please us with a mission of enhancing human health, exactly when do we want to them to make our appetites easy to fulfill, and when do we want them to help us be safer, healthier, and better environmental citizens?

When a driverless car – one of hundreds of millions likely to be on the world's roads within a few years – is faced with the choice of avoiding harm to one group of people (a cluster of pedestrians, say), or another (the people riding in the car), what values about whose life needs protecting most will be embedded into the computer logic making the faster-than-human decisions?

When our laptops, our phones, and little devices in our ears learn things about us that might impact our health, or cost us money, or keep us apart from people we care about – when do we want to know and when would we prefer not to?

And who else do we want to know about what we're doing and thinking and feeling? Sometimes we benefit from sharing – sometimes we benefit *greatly* – but when, exactly? When should strict privacy be the rule, rather than sharing information that can save a life, or help others, or conversely cause great distress because of the suffering of someone we care about?

Finally, with our digital systems generally built and run by some blend of corporations, governments, and creative individuals coming and going from large organizations that build and sell what they envision, when should these systems be 100% on the user's side instead of on the side of the systems' creators or managers, and how can we know?

These are the central questions of our immediate future, as important as questions of war and peace, and life and death. In fact, answers to these questions will shape our wars and our peace, our lives and our deaths, beginning now.

The political parallels
Principle-centered ethics is the only way to answer these questions, and others like them, in a way that digital systems can

make sense of and follow over years and decades, through the expected and the unexpected.

The first part of the process is the hardest part. That's the process of clarifying shared beliefs about fundamental questions like the relative value of human life, the limits of privacy, and the potentially different status of people and communities who might benefit from these systems.

Added to this most fundamental layer, we need to clarify the purpose of each system in question, and make sure that we define that purpose with boundaries that reflect the fundamental shared beliefs already clarified.

Clarified is a specific and important word here. If instead we talked about *establishing* shared beliefs our task would be vast and never-ending. So our best collective task is the effort identify where we already agree on the fundamental questions, code in and train in those points of agreement as boundaries for what digital systems must not do, and be sure to build our systems with options for amendment and adjustment as (we hope) consensus evolves and grows. A switch in a driverless vehicle that allows the operator to prioritize the duty of hospitality (privileging the lives of passengers over the lives of pedestrians when one or the other but not both face unavoidable harm) over the duty of protecting the public space over private space is one tool for this kind of adjustment on the fly, though a narrow one.

Consider again the precedents of the founding documents of the United States. Just as the Declaration of Independence establishes the principles of government not as arguments still open for debate but as bedrock shared values, the work of digital ethics begins with clarification, expression and rallying populations around similar fundamental values about what our systems are for, whom they serve, and what it means for them to do their jobs well.

This is the aspirational and inspirational beginning of digital ethics: planting a flag as the beginning of a process. We might say that all human life is equally precious and that this should be used as self-evident truth in all our systems. Some will reply that this is a lovely notion, but exceptions surely exist; that we can grind down too many functional and productive roles for digital systems by adhering to this absolute. And we need specific room to express our ideals and aspirations, and use them not as tools to *produce* code or law, but as tests of whether our code or our laws are as consistent as possible with our clearly stated ideal.

The step beyond principles – approaching the Constitution

It is the job of the U.S. Constitution, not the Declaration of Independence, to be the *producing* foundation of law in the United States. This longer, more complex and less-inspiring document is filled with compromises that can feel like retreats from principle. Yet by design these compromises attempt to capture as many of the ideals and aspirations of the Declaration as possible. In the case of the ambulance, an ethical guideline must do the work of delegating decision-making. But whose job – whose burden and privilege – is it to codify the principles that determine exceptions to the general rule of protecting water quality? Is it up to the company selling the software? The government agency operating that public-works system? How does the collective voice of the public speak to this practical instance of digital ethics?

Beyond clarifying basic principles we can mostly agree on, there is a vital task here to stand up and operate a system built on delegated responsibilities that affect an entire community.

In this case perhaps the rule for closing the road is enforced absolutely (the system fails to make ethical exceptions), or perhaps the system would not have been built and implemented at all among the babble of uncompromising interests, and the unprotected water made foul by a lack of coordinated leadership – a failure to address digital ethics.

To avoid these outcomes – one taking too little account of needed compromise, and the other expressing so much fearfulness of acting against consensus that it does not act at all – we need to understand the mechanisms of compromise in clarifying shared and unshared values and in making best-possible choices that represent the diversity of belief among us, while still allowing needed actions to unfold.

Two kinds of compromise

The first kind of compromise at the core of digital ethics is compromise *about* basic principles. The goal here is to establish principles that a broad range of people can support – not necessarily the exact principles each party would wish to support, which tend to be agreeable to a relatively small number of people, but principles close enough to their ideal but still appealing to others who, in turn, also accept compromise that leads them to agreement even though their ideal principles would likely be different.

From the digital-ethics perspective, an ideal compromise has two important qualities: it is meaningful, and it is accepted by the largest possible number of people even if it is not comprehensive or precisely what each party would prefer. I might prefer a principle that we write into the ethics layer of lots of systems that sounds like this: long-term protection of the natural environment is more important than short-term material gain. You might reply that protecting the natural environment is actually itself a material gain that we all share – cleaner water, more comfortable living conditions, and a range of additional material advantages are part of why we want healthy natural systems all around us. So we compromise on a principle, and both agree that we can write code based on the idea that we should favor long-term protection of the natural environment over short-term material gain for any one individual or small group of people. Elements of both positions are present, with some nuance that will require more thinking-through of these principles as they are applied.

The second kind of compromise at the heart of digital ethics is compromise in applying principles. The driverless-car example offers a good illustration. If we start with the principle that all human life is equally precious, and the driverless vehicle must in at least one instance choose a path that will lead to the death of one passenger or the death of one pedestrian, which path is best? Adhering strictly to the principle of all human life being equally precious creates a stalemate that puts the vehicle into a "do nothing" or "cease to operate" mode – in many ethically fraught situations, the worst of all options. In a case like this, the principle works as an ideal that we hope to reach to the fullest extent possible, even though we won't be able to meet its high standard in many cases. So we'll need a logic for compromise.

A Logic for compromise

A number of parties have an interest here, and for that reason among others a number of parties should have voice in how we make this compromise. That's why the second of the two documents in the second part of this book is all about what each of several different parties – technology creators, technology users, companies and governments, in particular – need to do to be properly engaged in the collective project of digital ethics.

The passenger and the pedestrian in this example certainly have a lot at stake. The auto manufacturer and owner stand at some risk as well, as do the general public and the government bodies that make and enforce laws and policies about the roads and about driverless cars.

Both the passenger and the pedestrian, presumably, want to live uninjured. Each has some practical moral interest in compromising the "all life is equally precious" principle to the point that he or she might live and escape injury. The auto manufacturer presumably wants the particular brand of automobile here to remain associated with safe driving – and because the passenger has likely invested more money in this specific vehicle than the pedestrian it has simply come upon, we can imagine a slightly greater interest in the auto manufacturer and in the owner of this vehicle valuing the life of the passenger more highly than the life of the pedestrian. The government bodies that make rules for the road and for automobiles have at least two special interests here: presumably they want the public to feel that the roads are safe, and presumably they want to be seen as doing a good job by the constituents who might (or might not) vote for them.

Because the pedestrian is in a weaker physical position to some degree – unprotected by the two-ton machine built in part to shield people as they travel – we might choose to exercise a bias to some degree for protecting the party who generally needs the most protection, in the tradition of American philosopher John Rawls. This does not invalidate the larger principle of all human life being equally precious, but reinforces its great importance as a constant point of reference, a stage or set of boundaries to limit the political interests and interests of the feeling of the crowd even while the full demands of the principle cannot be universally met.

The vital point is that the principle itself is the aspiration – the test we apply to the action in question. Among all options, we need digital systems to test each option against the principle and choose the one that comes closest while executing the task the system is built to execute, like moving a mechanical vehicle down a road.

This issue of which parties have how much voice in shaping compromise in applying principles is just as important as the principles themselves, because aspirational principles need a degree of idealism that must be balanced by the practical work of informed compromise, sticking as closely as possible to the principles all the while. Many commentators have noted that the Declaration of Independence is on its face too idealistic to actually *do* much. It is, in this sense, the poetry of the founding of the United States. The Constitution is the prose – the descriptions of what must get done, who must do it, and how. Yet the prose needs the poetry for these actions to remain full of purpose and to carry forward the ideals of shared human experience that only the poetry of pure principles can establish.

The digital ethics "Constitution" is about the getting of things done: in the actual building of systems, the selling of systems, the operating of systems and end-user's use of digital systems, what is the job within each of these tasks for people to do; what are the questions that they need to speak to as compromises are being made; what must they be aware of as they reach the inevitable compromises of coming close but generally net perfectly meeting the higher-level principles already agreed upon?

Back to the practical example: the digital-ethics declaration of principles might aid the driverless vehicle's system designers by beginning with the broad principle that all life is equally precious, with a bias toward protecting children more than adults. Then an act of counting seems called for: saving two pedestrians above one passenger seems to accord with this principle. But what about when there is one passenger and one pedestrian? If one is a child, the moral duty balances in that direction. If both are adults, then the system is at a stalemate unless an additional ethical weight is added.

The Constitution's best use is to identify who makes these decisions, rather than imposing the answers. In this case, it might indicate that the solution to a problem like forced prioritization of the value of human life in a critical situation is to be left to the makers of the vehicles, or to the makers of the systems that operate them, or to the citizens of the regions where these machines and this code is produced – but within certain limits. The Constitution might helpfully map out a hierarchy of exceptions to the general principle, including the notion that there is moral weight to the ancient ethic of hospitality, and thus allow the digital system operating the driverless vehicle to protect the passenger above the pedestrian.

The ringing phrase in the U.S. Constitution, "Congress shall make no law" has special importance as well. We might find that after establishing the ethical primacy of the equal value of human life, vehicle and systems makers create options to prioritize the pedestrian or the passenger as they feel best represents the values of the people they most directly serve – or the values of their cultures they find themselves embedded in – but within some explicit boundaries. Perhaps we land on a notion like "systems makers shall make no systems, and operators shall operate no systems, that result in the general privileging of human life based on the identity of any individual or group." This is a step more practical and more mindful of compromise and complexity than the easier Declaration-like language of all human life being equally precious, though it attempts to preserve its spirit. At the same time, it anticipates evolving circumstance and shifting local ethical norms and allows change *within limits*. The human hand and the human voice must set those limits with an awareness of the dangers of automation and computational control of our culture, and the strongly positive effects that the evolution of smart machines and systems have had over the last three hundred years.

This is the tricky business of compromise – detailed, shifting, and rooted in clarifying whose voice is to be heard.

A case for optimism

Uninformed optimism is foolish and dangerous, of course. But so is a pessimism based on fear rather than careful observation. This tension between optimism and pessimism is especially important as we consider how much our work in digital ethics needs to be fundamentally defensive, and how much that work ought to recognize and amplify the shared human progress that can and should flow from increasingly speedy advancement of technology.

Are we reaping these benefits yet? In many ways, yes. Yet the doubters have reason to doubt. Humans are indeed becoming in some ways obsolete, and our abilities to even understand the world we live in are declining relative to what computers know. Jobs are already being lost. We already put our lives in the hands of computer-driven systems every time we get on an elevator, fly in an airplane or seek treatment in a hospital. And coming soon: every time we're in a car. Every time we drink water. Even every time we breathe. The very idea that *computers can and will run the world* makes some of us nervous – more nervous at this particular point in our history than ever before, and rightly so.

On the other hand, we're already vastly better at connecting with people we care about even if they're on the other side of the world than we've ever been before. We're better than ever in growing more food and in learning about the lives and politics and arts and ideas of people all across the planet. The poorest billions on our planet are getting steadily less poor; a larger number of people living on the planet are going to school and learning to read and write than ever in human history. New chips and computers and systems help make this all happen, as advances in communications, in medicine, and food production over the past few decades make undeniably clear.

In the already-wealthy parts of the world, we have dinner in New York with dear friends in India by placing an open laptop on the table and Skyping them in; pediatric heart surgeons add decades of life for children born with heart anomalies by practicing their operations on exact replicas of their small patients' hearts, 3D-printed specifically for that purpose; our smartphones guide us from neighborhood to neighborhood and tell us where we can find the nearest friend, restaurant or hospital in places we know nothing about. Already. And in the less-wealthy parts of the world, we can count the gains among the poorest in the billions – billions of people with higher incomes, longer lives, and a greater knowledge of the world.

But as we take the next quantum step up the ladder of power we give to automated systems of all kinds, we face one paramount challenge: to keep the human touch. More specifically, we must clarify our human beliefs and write them into the computer code that runs our systems, so that these beliefs set limits to what computers do.

The human touch in the hardest moments

This human touch is particularly important when there's no good option in front of us and we have to decide quickly which among a set of bad options is the least worst. That's when our ethics really come to play, when we decide in a split second what the right thing to do actually is, when "right" in effect means "least wrong."

Some people think about and talk about these kinds of choices often and might have a clear plan for how to act under stress. Others belong to communities that answer a lot of these questions in advance in structured ways, like military outfits with clear rules of engagement or religious groups that lay out when to it's OK to do harm for the sake of a greater good. But most of us don't think and talk about these issues much and discover what we believe in the split second when we make the hardest decisions.

That might come when you are alone in your home and you see someone in great need – someone hurt, someone terribly cold – out your window. You either go out to help this person or you don't. You either take this person in or you don't. You either call on others to come and help or you don't. Your ethics become clear in your actions – whether you've thought about them clearly or not. And now is the time to build human values into digital systems, so that they can make similar choices. But we need to be explicit about the principles they will be called to act upon – there's no room for the split-second judgment in our digital systems without clear boundaries wired into their instruction-set.

"The Box" – Let computers out, or keep them in?

Can computers think "out of the box" enough to latch on to other systems, or start doing things we haven't directly told them to do? Without a doubt they can. An important part of this process is the creation of what programmers call self-governing algorithms that learn so much faster than people do that they are trusted to expand the number and kinds of tasks they perform, given that they know a lot more, a lot faster than we do.

It's often tempting to let self-governing algorithms be *really* self-governing – to build in few or no restrictions on what they might think up as new tasks for themselves, because we can potentially benefit from their speed and power, and because many of those new tasks are likely to provide humans with greater benefits from the work of these systems. This temptation is like the temptation to allow a strong-man dictator free rein as a political leader. These people tell us that they'll do such a good job, and make such good decisions that we don't have to worry as much as we've been worrying – we don't even have to vote. Trust them – they're so smart and so strong, we can relax. And if voting's still important for us to feel good, we can all run out and vote for the new boss with no real opposition presented or allowed to speak up. It's certainly efficient, and we do seem to have an instinct to want to worry less and let the complicated matters that we've thought perhaps too little about be decided by a trusted leader.

Computers can fill the same role. They're so smart! They can absorb so much data! They've done a pretty good job so far. . . so let's let them make more decisions for us. The algorithms at the heart of how they evaluate the data that we feed them in rapidly growing quantities can help them not only solve the problems we've built them to solve, but find new problems and solve them even before we notice them. Think about the system that manages the mix of fuel and air under the hood of your car. Two generations ago, drivers needed to problem-solve when their engines knocked or sputtered, and manually expand or limit the airflow in the engine through a little plunger in the dashboard (the "choke"). Today, not only do our more-automated engines manage the mix, most drivers don't even know that level of O2 in the fuel is a potential problem.

The system solves the potential problem before the user even knows it's a problem – and in the process, the user loses a sense of how the engine really works and what's happening in the complicated combination of moving pieces that makes the car go. If we allow the engine-management system more freedom to identify and solve more problems, our jobs as drivers get easier and easier – and eventually we won't even have to drive at all. But in the process, we must limit the types of problems that we want these systems to solve, and the range of solutions that are acceptable. As we stop thinking of the engine's task as running at certain speed, with the right flow of fuel and air and the right pressure in the tires and instead think only at the more abstract level of the engine's job as getting the car and its passengers from here to there, we need to emphasize that some speeds are too fast under any conditions, and some trade-offs not acceptable even if they get us to our destination faster. Similarly, we don't want to lose the skills and knowledge about how our government works in exchange for greater efficiency, more order and less messiness in our public spaces. The inefficiencies of human experience and ideas, of human values and surprising difference among people, have a value that rises above the explicit tasks of government.

The democratic process is built to be inefficient, so that efficiency will never shut down the constant re-imagining of what is acceptable and valuable in shared public life. It's a process built for revision and amendment of all kinds, with the inefficiencies of values and reflection balancing the prospect of practical gain. More efficiency is of course tempting. Cleaner streets are tempting. Trains running on time are tempting. Not having to hear opposing ideas or troubling complications or bad news is tempting. But there is a greater good in the democratic process – the engagement and importance of each individual, and the fact of the process itself, are goals just as vital as any outcomes and outputs. So too with digital systems.

Three things that creators and operators of digital systems have to do

In response to these temptations then, people and organizations who create and operate automated systems of all kinds have to do three things, starting now, as necessary parts of doing their work in every case.

First, systems and algorithms must operate with independence in many ways, but they cannot set their own course. <u>Purpose and function are fundamentally human decisions and human designs</u>. That is our irreducible, unavoidable and essential role in the creation of technology, no matter how smart that technology.

Second, ethics – the statement and application of purpose and values – must be the first layer of every technology stack. A foundation of ethical principles and limits is the bedrock on which the further layers of a technology stack must stand. Without that foundation underlying the stack, the stack cannot be stable for the long term.

Third, our technology endeavors will always need to address the three core ethical questions mentioned in the introduction, as part (but not all) of that foundation of the technology stack:

How much should privacy matter?

How important is the ownership of intellectual property, and what ownership and use right need to be built into each system?

How should we evaluate and work for fairness and equity in how benefits of technology accrue?

The second half of this book gets more specific about commitments that we can all make to address these questions. But the answers matter little until we can feel the importance of the questions.

In digital systems, we can tell how important a question is by the position it holds in the hierarchy of the layers of code that come together – the stack – to make the functioning whole.

Getting digital ethics right means getting an ethics layer in the proper place in the stack. And that place is the primary position, because purpose and direction are the province of humans, not machines. We need to build our systems with human-crafted and human-observed purpose as the most fundamental layer, every time.

Chapter Two

The Stack (How to Run Toward the Danger)

A police officer walking down a city street hears gunshots ring out. They're loud and close by. People strolling by stop, look around, hear more shots and instinctively scramble toward nearby storefronts and apartment-building doorways. They're confused and afraid. More gunshots. The officer is already running – *toward* the sound, *toward* danger.

Why? The answer is not complicated. The police officer is committed and trained. She's made a commitment to protect others by engaging sources of danger, to make the world safer than it would be if threats go unchecked.

The officer is acting in a moment of crisis – a moment when an unplanned event forces an instantaneous decision to go this way or that – based on three critical factors:

1. **A clear general principal about the right thing to do** in the face of danger;

2. **Reflection on that general principle often and openly**, though conversations with other officers, friends, and family. The motto "to protect and to serve" is written on the door of the police car she gets in and out of a dozen times a day. At the morning briefing she attends every day at her station house, she hears her supervisor remind all her colleagues that "our job is to make this city safer." The general principle is recalled and recited often.

3. **Practice acting on that general principal** so that in the moment of crisis, following that principle is a matter of habit as well as choice.

Of course, police organizations in some cities and towns are better at this than others. The best hit all three of these critical factors every day: clarity about their values; frequent and open reflection on those values; and practice acting on them.

The high school that starts every day with a gathering of students and faculty to welcome everyone to another day at the school, and to give the principal the chance to repeat the school's core values, is acting in a similar way. The principal says to the community, every morning, that "we come here every day for all of us to learn, and for all of us to teach, and for all of us to show concern and compassion to each other." A clear set of principles is laid out, the community reflects on them openly and often, beginning with that morning meeting, and – we hope – opportunities for everyone in the building to practice teaching, learning, concern and compassion are cultivated.

The foundation that states its core values in every document it publishes and every message it promotes in the media – "We believe that all human live is equally precious, and invest to create more health and opportunity for the greatest number of people we can" – does the same.

The results of these three critical practices is that the individuals who work and study in these communities develop the instincts to act in accordance with a set of core values. Especially in moments of crisis, and especially when facing new and hard-to-anticipate circumstances, these values point to the path of action in line with what these communities believe. And they help make clear *what these institutions are for* in two senses: what they're built to do, and what they advocate.

These three vital practices – clarifying fundamental values, reflecting on them openly and often, and practicing how we can act based on them – form the base for ethical behavior in organizations like public-service and government agencies, schools, and most certainly businesses.

Back to the stack

In a pure technology setting, stacks contain countless abstraction layers. The bottom layer of a stack might be zeroes and ones (or quantum qubits in the not-too-distant future), building all the way up to a nice app that helps you send flowers to your mother on your smartphone or launch a satellite.

The stack begins with tools that can be applied and combined in ways that their makers don't necessarily anticipate. Each layer creates a broad range of possibilities for what can come next, but each layer also sets parameters and establishes some limits. It's a lot like starting with an undefined mass of water – not rock, not sand, not sunlight, but water – and then adding a layer to put some borders and edges around that water to make it a defined mass, and then a layer that subdivides that contained amount of water so that it can be picked up and carried around, and then a layer that creates a handle and a spout giving the water some direction, and then a layer that lets you tip and point the water carrier so you can water flowers, or direct a stream of water to a steam generator, or take a shower.

The stack only works because of that first layer – it creates the possibilities by defining what you're working with – and layers on top give the enormous potential some shape and direction. Computing systems that do things like run the power system in a big city or control the flight patterns of planes in the sky, all run based on a stack of computing technologies that interrelate and build on that first layer.

We can see "first layer" statements of values in some of the texts that underlie our governments and religions – and that's no accident. The Declaration of Independence includes this powerful statement, which is indeed a fundamental statement of values that other, more refined and practical ideas and actions build from: "We hold these truths to be self-evident: that all men are created equal, that they are endowed by their Creator with certain unalienable Rights, that among these are Life, Liberty and the pursuit of Happiness."

Homer's epic poems tell us that hospitality for travelers is among the highest forms of goodness – we must treat the stranger with generosity. The Hebrew Bible gives us the ten commandments. The Quran instructs that all wealth is meant for sharing – that people with means are obligated to share with those who have less. These are fundamental values, and form foundational layers in the stacks of communities and civilizations that build upon them.

Digital ethics works on a stack as well

Digital ethics works on a stack as well, and the first layer has to be built of core, clear values. Let's say we've created a set of digital technologies that autonomously produce fine furniture at a modest price. Robots take lumber from a warehouse, deliver that lumber to workstations where more finely-tuned robots cut the lumber to size, build frames, add stuffing, sew on fine finishing cloth, and wrap chairs and sofas into shipping boxes that emerge from loading docks to be picked up by automated loaders that place the boxes onto flatbed train cars that carry them to stores in nearby cities. All automated, all governed by software.

The first layer of the digital-ethics stack in this case needs to be a crisply expressed basic principle that will govern choices the system has to make in the event of shortages, accidents, unanticipated customer demands, power failures, rising or falling prices, and any number of other possible challenges this digital furniture-making system might face.

In this case, the humans who think up, invest in, and plan the system spend a few weeks talking about what they really want this system for, and what they want to *be* for. They come up with this: technology should deliver high-quality furniture to large numbers of ordinary people at low prices to make their lives more comfortable. Each of the key ideas here will help set parameters on what this system will and will not do, and will create a model for making seemingly hard choices in unexpected circumstances.

If, for example, supplies of quality wood suitable for the system's manufacturing needs become restricted, the clear options will be to sacrifice quality but continue to produce a large amount of furniture at low prices; produce less furniture for a small number of customers at a low cost and high quality; or use a non-wood alternative to keep quality high, volume high, and prices low.

The carefully crafted first layer of the stack is helpful here. It limits the range of quality the system is built for: high quality. It limits the range of volume: high volume. It limits the range of cost: low cost. But it does not limit the use of materials: using non-wood alternatives is fine according to that first-layer set of principles.

Because the fundamental purpose of the digital system – the answer to the fundamental question *what is it for?* – is well thought-through and clearly expressed in this case, the answer to the momentary question *what should the system do now?* is easy to derive by applying the purpose to the facts of the situation. Rather than programming in a specific instruction of how to deal with this kind of challenge, the principle plays an efficient role, itself working like an algorithm: it generates an answer to the question of the moment based on the principle, rather than pulling an answer from a catalogue of answers that includes an entry that fits the moment. The efficiency here, and the ability of a system built with a clear ethics layer in place, is not the only virtue to note. This model of using a principle to generate answers and limits even for unimagined situations equips digital systems to be self-limiting – to avoid the temptation to do more and more, and control more and more of human experience.

For a second example of a strong ethics layer in a stack, take the case of the Gates Foundation. This is part of what the opening screen at www.gatesfoundation.org looked like in 2018:

This principle that all lives have equal value is not a conclusion based on evidence, or the result of a scientific investigation. It is a statement of belief. Many alternatives are possible (for example, "The lives of my children are more valuable, in my eyes, than anyone else's"), but this is the position that the Gates Foundation founders and stewards have chosen. It has the irreducible quality of a faith statement – you can't refute it or correct it, though you can choose to disagree and adopt an alternative position. But it's a foundation in the best sense – the first level in an ethics stack that sets boundaries and direction and can be applied as a compass in uncertain and unanticipated situations.

Any organization with this principle in place should have no trouble choosing between investing in creating new vaccines for diseases that affect two or three million people and investing in sanitation, clean-water projects or improving roads in areas where large populations are isolated and prone to sickness. Across the world, about four million children die every year from dehydration. No vaccine is needed to cure dehydration – water does the job, though basic sanitation is vital to ensure that clean water stays clean and dirty water gets clean enough to drink safely. In most cases, the effect of investing in roads and sanitation will make access to clean water far easier for far more people and help improve and even save more lives than more sophisticated efforts to cure diseases spread by dirty water. The principle applied to the circumstance offers a prompt to one kind of action more than another: spend the first dollar to foster greater access to cleaner water. Spend the dollar to cure disease

– rather than prevent disease – only after access to clean water has reached a critical mass.

Does this sound too simple? Indeed, in many ways it is, and the real-life Gates Foundation does invest in some vaccine development, but not much by relative standards, and it does fund some programs in part because they help to open doors to other opportunities to do more good, more efficiently. Still, the guiding principle is identifiable in the overall picture of where the foundation spends its money. It is the kind of principle that has some exceptions – more about handling exceptions to ethical principles later in the book – but it does its work as a guiding principle quite well, not dictating what the Gates Foundation does by clearly setting a standard and aspiration that both individual efforts and the cumulative whole of the foundation's activities should be, and are, measured against.

Practical guidance from the ethics layer

An ethics layer beginning with a fundamental principle like "all human life is equally precious" can support any number of different applications that sit on top of it. We can use this base to build a stack that leads to an important guiding principle for driverless cars for example. Faced with the plausible situation that you or I – or our children – are sitting in a driverless car when circumstances reduce all of our options either to driving off a cliff or slamming into a couple of charming neighbors, this principle can guide us to a mathematical consideration.

Since all human life is equally precious, it matters not at all whether the neighbors in front of us about to have a deadly run in with our driverless car are young or old, decent or cruel, happy or sad. We know – our driverless-car's operating system knows – that all human life is equally precious. And so if there are three neighbors facing the onrushing vehicle and only one or two of us strapped in to our automotive decision-making machine, we and not our neighbors ought to go over the cliff. Yet if the numbers are different – if there is only one of them compared to two of us – it's a bad day for the neighbor.

A different stack built to shape the ethical practices of a driverless car might have begun with a first layer including the idea that we have a duty to protect the people closest to us by blood and circumstance more than others. If this is the stack we build, then our children will trump our neighbors, our neighbors will trump our distant acquaintances, and so on. The digital ethics stack has what it needs – a principle that the machine can apply quickly to a circumstance in which some people would likely make decisions based on similar principles not yet put into words or clear thoughts, but felt. A driverless-vehicle operating system without that ethics layer in place might, faced with a dramatic choice like this, do nothing and simply continue on its path – over the cliff if doing nothing were to lead to that; into the neighbors if doing nothing might lead to *that*; or into the neighbors and then over the cliff if the specifics of the road make that the path of least resistance. Without the ethics layer, our digital systems and machines become like the driver frozen at the wheel, missing the chance to make a choice that can save this life over that life, or these lives over those lives. It commits the ethical errors of the blind machine, a chilling prospect, rather than the expressing the inevitably flawed but humane impulses of our human culture, our human hopes and what might pass for our human wisdom when faced with the hardest decisions. In the world of artificial intelligence, more and more data, more and more experiences based on more and more incidents will allow the work of our digital systems to continuously improve and adjust. But our fundamental principles should not be so easily altered. Billions of instance-based ethical decisions should not change those principles – that work belongs exclusively to humans.

Seeing the stack in the non-digital world

Consider, again, a police officer. The officer is trained in a set of values as well as in a set of skills. The training begins with answers to questions like, "What is this job about?" and "What are we here for?" There's actually been a lively debate in law enforcement for many years over what some call the "guardian versus warrior" choice. That is, are police guardians who should be thinking at all times about how to make people's lives safer and better, or are they warriors, thinking at all times about defeating an enemy? Certainly, the best police departments manage to play both roles, but one almost always predominates. This choice – warrior or guardian – is about the first layer of the ethics stack: If the fundamental principle at the first layer is "The job of the police is to be guardians of all the people in our town," then the policing concept of "de-escalation" becomes far more important than the idea of defeating the bad guys. Police will frame an unfolding conflict less in terms of the "bad actor" and more in terms of the collective interests of everyone involved with a focus on preventing harm and reducing head-to-head tests of power. In contrast, if the first layer of the stack is "The job of the police is to control and defeat those would do harm in our communities," then policing will likely become more confrontational.

Here's what it looks like when police are guardians first. In April of 2015, four young police officers from Sweden were visiting New York City, and found themselves in the middle of an ongoing crime scene.

From the *NY Post*, April 22, 2015:
"The Scandinavian patrolmen said they were on their way to see a performance of Les Miserables when they heard the operator of their uptown 6 train yell frantically over the intercom: "Are there any police officers on the train?!"
"We thought maybe someone needed help," said Samuel Kvarzell, 25, a rookie with the Stockholm Police Department.
When they made their way to the front of the train, they saw one homeless man beating another senseless, as terrified straphangers fled the car into the Bleeker Street station.
"One of the guys was on top of the other guys, so we separated them," said 25-year-old Markus Asberg.

And, fitting our digital age, several people caught much of the action on their cellphones. A video of the four Swedish policemen on holiday shows them restraining and comforting both combatants – two homeless African-American men, both on the dirty floor of the subway car, one raving with two of the off-duty officers restraining him, the other sitting calmly.

The contrast to videos of police killings of African-American men circulating with particular poignancy at the time – just as the Black Lives Matters movement was gaining traction – was striking. Social media lit up with re-posts of the video and comments like "My God, they're so gentle." That gentleness was no accident; it was what happens when a clearly articulated general principle is laid as the foundation to support more specific actions as a stack is built. In this case, the stack is a set of ideas, trainings and reinforcing experiences that produce police work. Because these Swedish police have been deeply and regularly exposed to the foundational idea at the base of this stack – that their work is to produce more safety and peace for everyone in the communities where they work – their daily practices, including how they run toward a problem instead of away from it, how they talk with and engage angry and violent people, and how they restrain and talk with people who have done violence and seem ready to do it again, are all shaped by that first layer in the stack. The ethics layer in that stack not only helps them know what to do and how to do it, it also rules out things they might do but won't, because those things have too much potential to violate the foundational principle about helping everyone. They did not scream at the fighting men; they did not beat them; they did not treat them like enemies. Instead they subdued them, soothed them, and expressed the values their system is built upon.

Knowing more means facing harder choices

Just as a good computer database tool can have layers built up on it that make databases of ice cream flavors or databases of political prisoners, so the stack that begins with *all human life is equally precious* can have layers built upon it that point toward on-the-fly driverless decisions of life or death, or the practices of police officers, or the policies about the distribution of scarce resources by a government, or the structure of a healthcare plan for a family, a company, or a nation. Each would be a different stack built on that same initial layer.

That first layer – the ethics layers – has decisive impact. If we say, "all human life is equally precious, except that children have a special status and should be protected before adults are protected," then our driverless cars will protect children before they protect adults. Children will be prioritized when we ration scarce resources that will sustain life in life-threatening circumstances. Healthcare plans will start by ensuring services that make a difference in the health of children before they provide services that make a difference for older people.

And in the digital age we are just now beginning, these things will be happening in an automated fashion – not only because a driverless vehicle will at times be making instant and automated life-and-death decisions, but because digital systems will be allocating and distributing food in times of crisis and scheduling our healthcare visits, allocating access to medicines, and reacting to news of emerging medical needs revealed through wireless monitors embedded in our phones, our vehicles or even our clothing. Today, with relatively little assistance from digital systems in planning and delivering support for people in need, it's wonderful to see someone alone who suddenly needs life-saving help actually get that help – through a neighbor's efforts to check in, perhaps from a phone call to the emergency service, perhaps from passer-by on a public street noticing the need. The fact that so many people with great need for help are less visible than others – that their need makes it harder for them to communicate, to be out in the world, to clearly express their struggles – actually suggests that in a more digitally-networked world we'll be better able to know who needs help and reach out in more effective and systematic ways. Consider the example of the elderly person living along and suffering a mild stroke or a sudden high fever. A hundred years ago, what could that person do to call for help? Short of the physical strength to get up, walk out the door and find a neighbor or passer-by, not much. With the advent of the telephone, the physical task gets simpler: just get to the phone, recall a number, dial it and tell the person on the other end "I need help," but even that might be too much to ask. With a cell phone at hand, the bar is even lower but perhaps still too high. If there's a digital device like Alexa in the house, it's easier still – but still the clear voice and the coherent sentence are required.

Yet we're not that far from that smart-phone or smart watch or digital toothbrush or tiny peel-off monitoring patch knowing that something terrible has happened – or is about to – and reaching out to summon aid without being asked. Our doctors and EMTs and firefighters and neighbors and town leaders will be able to see the need and help can be on the way. *Their* digital systems will check in, dispatch, order more emergency supplies, create monitoring schedules, and make a systematic difference in the lives, potentially, of everyone who needs help. Yet when we know all the better *just how many* people really need help, we will have to tell our digital systems how to choose who gets help first, and in the inevitable sad cases, who gets left out when resources for help are used up at a given time in a given place.

The not-knowing of a lower-tech world in some ways has shielded us from making difficult decisions like this. As our digital systems help us know more and more, and demand of us that we answer questions like *who gets help first?* in advance, based on carefully thought-out principles, getting those principles right is all the more important.

Who controls the ethics layer? Lessons from the Milgram Experiment

In 1961, the United States was only beginning to come to terms with the moral legacy of World War II and the Holocaust. In fact, "the Holocaust" was a phrase not yet much used, and even the refugees from Hitler's extermination camps who found safe haven in the U.S. often preferred to avoid talking about that searing chapter of history as they rebuilt their lives. That year, though, a German officer responsible for the deportation of millions of Jewish men, women and children to death camps was put on trial in Israel after he had been snatched up by Israeli secret agents in Argentina, while living under an assumed name as a manager at an automobile plant. Adolph Eichmann was tried and found guilty of crimes against humanity, and crimes against the Jewish people, and executed.

His trial was front-page news across the globe, certainly including the United States where all the major daily papers covered the turning of the judicial wheel, and the *New Yorker* featured the philosophical reporting of Hannah Arendt, herself a refugee from Hitler's Germany. Her work was later collected in the groundbreaking book *Eichmann in Jerusalem*, which introduced her phrase "the banality of evil" to the world. Within a few years, conversations about the Holocaust and how seemingly decent, educated people by the millions came to countenance and even assist in the killing of their neighbors, also by the millions, would become more common.

In that same year of Eichmann's trial, a social psychologist at Yale University named Stanley Milgram began a series of experiments about obedience to authority to test whether ordinary American men would follow orders to do horrible things, even to the point of killing others, if they felt compelled to by authority figures – not at the barrel of the gun, but at the insistence of someone seemingly in charge.

While the Milgram experiments became something of an industry spanning decades and continents in all their variations, the core experiment – conducted in the basement of a Yale University building in New Haven, Connecticut in 1961 – was simple. Two men worked as experimenters, while one test subject at a time paid the then-princely sum of $4 an hour to help further science was brought into the lab. One of the two experiment staff members wore a white lab coat and presided as the scientific expert. He was clearly in charge. Another staffer posed as the "learner," whose job it was to remember random word pairings that the experimental subject would read out. The subject would then test the success of the learner's attempts to hold the word pairs in mind. The subject – playing this role of reading off word pairs and then testing the learner's memory – was called "teacher" during the experiment.

To kick things off the man in the lab coat, the learner, and the teacher – actually the experiment's subject though he thought he was only helping out by testing the learner's memory – all shook hands and established a friendly tone. The teacher sat in front of a long 1950's style electronic consul with an impressive row of switches each in the off position. The switches were each labeled in ascending order with a number of volts, the last few marked with the word "danger" and a row of exclamation points.

Before the learner would go into a small adjacent room, supposedly to be hooked up to a machine controlled by that long electronic consul, the teacher drilled the learner in a series of random word pairs, like "red-ball" and "window-book." Each word supplied by the teacher needed to trigger the right reply by the learner, or else the teacher would push down one of those switches and the learner would receive a shock. And the next time the learner failed to demonstrate proper learning, another shock was delivered by the teacher, one step higher in voltage.

In fact, the learner wasn't hooked up to any machines at all, and all eyes were on the teacher – he was not there to help run an experiment on anyone else. The experiment was being run on him, to see the extent to which he would inflict harm on an innocent, pleasant person simply because a figure of authority (in this case, a man in a lab coat) told him to. Most of the "teachers" hesitated to inflict harm and tried to be excused from the task early in the experiment's unfolding as the learner was heard to say "ouch" and "hey, that hurts."

The lab-coated man said plainly that the experiment needed to continue. It the teacher protested further the man in the lab coat became a bit insistent that the experiment required that teacher continue. Finally, if the teacher continued to protest – and the original film of the experiment captures many who do protest, and several clearly in a kind of moral pain as they face their task and the choice to continue – the teacher was told simply that he had no choice but to continue.

Watching the film of that early version of the experiment today, one particular subject stands out. When told "You must continue. You have no choice," this man immediately replied, "Yes I do have a choice." This was a man who had a sense of the stack his actions were a part of, and that there was an ethical layer at the bottom that ensured he had a choice. His courage and insight rely entirely on his sense that a fundamental value is at play. The moral system he feels a part of makes possible an assertion of individual conscience and will – as well as a principle about not harming the innocent without good cause that was clearly (to this teacher at least) in play. This was precisely the kind of sensibility that Milgram was probing for, or for the lack of.

This man was drawing rules about how to live from that first layer in the stack of his own "technology." He felt the first principles – principles like recognizing the humanity of others – and then more refined layers in the stack, like treating others with respect, to finally explain that the moral consciousness he brought with him to that small social-science lab at a university would not allow him to hurt an innocent man beyond a certain point. No experiment could require that, because checking the higher-level element of the stack (the experiment) against the foundational level (perhaps "do not harm the innocent when there is any alternative"), a contradiction means you stop, you act within the bounds of your first layer and not beyond them.

The experiment has been broadly seen to demonstrate how relatively little it takes for ordinary people to do things they find wrong – to violate the moral principles they likely hold, or think they hold. The film of the experiment makes clear that these are not people inflicting harm because they think it's a good idea. And they are not passive even as they do what they are told. They know they're causing harm and that causing harm is wrong, but the force of the man in the lab coat's authority pushes most to do what they feel is wrong, squirming, sometimes begging for release as they hear the screams through the wall.

They clearly seem to feel that they are part of a stack that should rule out the kind of thing they are doing. The ascending layers of values and rules and capabilities for how to live the lives they live simply should not allow what they are living through. It's almost as though the stack they thought they were a part of has broken – or that the first layer of that stack turns out not to be what they thought it was.

And in fact the man in the lab coat is in effect telling these "teachers" that they are part of a different stack, and he's explaining how that one works, what the rules of this slightly different stack now in force actually are. The man in the lab coat knows the rules. He explains them, and – in theory – the teachers have no choice but to follow along. Except for those who seem to understand, at least for a moment, that they do have a choice – that they can remain connected to the stack they believe in, grounded by moral principles they refuse to abandon.

"Of course I have a choice," the brave, principled teacher tells the man in the lab coat, and one might cheer. Though after his flicker of clarity and conscience, he goes back to the consul, and back to running up the electric shocks.

Chapter Three

Digging Up Our Buried Values:
Revisiting the Man and the Dog (Adding Karl
Marx, and a Woman in Chicago).

Let's return to the story of the man and the dog in the factory of the future, mentioned earlier in this book. The man's job, Warren Bennis explained some time ago, will be to feed the dog. The dog's job will be to keep the man from interfering with the equipment. This charming but chilling story about work and automation is also a story about human ideas, and how our society continues to evolve in ways predicted by thinkers ranging from Aristotle to Karl Marx, who envisioned a world of warm community, art, leisure and greater social engagement for every worker once communist revolution redistributed labor and its rewards, allowing the working man and woman to share more fairly in the value of their labor and the rewards of technology.

Marx foretold a world with more productivity from the work of men women because of technology, and therefore – in a system more fair than the brutal capitalism of nineteenth-century Europe – more leisure time, more money and more autonomy for the worker, allowing humane values to flourish. Certainly, those acting in Marx's name failed in the 20th and 21st centuries (so far at least) to bring about anything remotely like the flourishing of the common good he predicted once the workers seized the means of production and administered the state for the sake of the New Man through a government of advocates and wise heads. The advancement of digital technologies seems in the last three or four decades to have moved forward this vision of a work world infused more by caring, education, art, health and wellbeing than by building, packing and selling things– aided by government engagement in supporting the common good, engagement helped along to at least some degree by Marxist labor reforms and movements, along with religious communalism and utilitarian concerns for a stable, healthy and peaceful proletariat by the ownership class. It's a complicated story, but with a trajectory: toward less physical creation of things, and more shaping, relating, explaining and adding human stories to things.

Consider, for example, that the large majority of labor two hundred years ago was spent in physically making food – helping plants grow and harvesting them; raising animals for meat and managing their growth and their material journey from living creatures to food ready to cook.

The 1800 U.S. Census reported a total workforce in the young nation of 1.9 million (about 1.4 million were free people; more than 530,000 were slaves). Roughly 1.4 million of these nearly 2 million workers worked in agriculture, and another 5,000 worked as fishermen. Hundreds of thousands more were famers in addition to other full-time jobs recorded in that year's decennial census, including many of the 40,000 workers on ocean-going ships, and many of the 40,000 free people working as domestic servants (these were the two most popular occupations other than "agriculture," the category claiming 1.4 million).

One hundred years later, with 29 million workers in the US, 11 million were full-time agriculture workers, and many in other occupations, including the 1.8 million in domestic services (the second-most popular category after agriculture), still tended to home farms and gardens vital to their diets. By 2010, with 129 million full-time workers in the U.S., fewer than one million worked in "farming, fishing, or forestry," while 6.2 million worked in "food preparation and serving." For every farmer, there are now at least six waiters and food-prep workers – not making the food, but making the food look and taste good, moving the food the few feet from the grill to the table, and in some cases telling stories about it to make the food more meaningful to the people who eat it.

Ye Olde TED Talk

Imagine for a moment a story told about two hundred years ago, not about the factory of the future but about the farm of the twenty-first century. Tell the small landholder of 1820 that in the future, the job held by more men and women than any other– the small-scale farmer – would be almost obsolete. No more jobs for more than 90% of the world's agricultural workforce, the most visible and celebrated application of human effort and intelligence in just about every civilization across the world at that moment in the 19th-century, as it had been for centuries. Imagine the questions: if you don't farm, how will you eat? If only a few people farm, how could there possibly be enough food for all? If a large quantity of food will be produced by only a very small number of people, what motive could there be for those few to share the food with the rest of us – the obsolete farmers no longer needed for their labor?

And now, two hundred years later, fewer than a million Americans are farmers out of a total population approaching 350 million and a workforce of about 160 million. The shift from 90% of Americans working as farmers to fewer than 2% is remarkable – and the technology futurist of the age of the American Revolution might have built a nice profile on the big-ideas circuit with sheepskin posters telling the story of the decline of the farming economy.

Call it the original TED talk: What shall we do when new technologies wipe away the work of 82% of our workforce?

Well, we now know the answer: we'll live a lot longer, spend a vastly greater amount of our time learning and entertaining ourselves, and live far more comfortable lives. Things will turn out pretty well, by the sparse standards of the eighteenth-century farm. Even – and perhaps especially – among the poorest and least skilled progress will be dramatic: a tiny fraction of infant and childhood deaths (in 1800, 43% of children died before their fifth birthday according to the demographers at Gapminder); much longer life expectancies (estimates for average adult life expectancy in the U.S. as late as 1900 range from about 45 to 50); and overall health, even among the poorest in industrialized nations, is strikingly better. And even beyond the lucky minority who live in the wealthier nations, the quality-of-life measures that predict an even better future have begun rising dramatically, with more than half the world's population receiving enough schooling today to be able to read and write, according to the United Nations.

On the other hand, while the majority of people in the world are indeed eating better and living longer, most of us are not nearly as connected to the practices and systems that make our food as our grandparents were. We don't know what farmers once knew, and we don't have the kind of democratic engagement with the making of food that shaped our society when the average working man or woman was a farmer in one way or another. There is a cultural loss here, and a potential shift away from intuitive understanding of how the basic elements of sustaining life work – and perhaps what they work *for*. We live much better material lives than our great-great-grandparents did, but most likely we understand less about where our food comes from, how the elements of our natural environment are connected and how the machines we might encounter on any given day – whether they be sailboats, steam-engines, riverside mills or jet planes – work.

And now that we're at the beginning of an enormous technological shift that may come close to zeroing out the jobs of American truck drivers (about three and a half million Americans make a living driving trucks today, but in the driverless-vehicle era that number will likely plummet), taking most of the humans out of the daily decision-making loop for running our airports, our electric plants, our factories, and even our financial institutions, we're anxious about the future of work.

Here history tells a hopeful story. Two hundred years of precedent suggest a social bias toward sharing the benefits of automation. Even as human labor has become less and less necessary to make food, we've found ways to keep each other fed (mostly). Chances are pretty high that the same collective interests and human values that have prevented us from starving the obsolete farmers among us (that's most of us), or pushing them off a very big cliff, we'll figure out how to fill our days in rewarding ways after the jobs most of us work today are further automated, and "work" continues to shift from physical creation of stuff to the learning and sharing of ideas, entertainments, and social experiences.

Consider this often-quoted remark from U.S. President John Adams:

I must study politics and war that my sons may have liberty to study mathematics and philosophy. My sons ought to study mathematics and philosophy, geography, natural history, naval architecture, navigation, commerce, and agriculture, in order to give their children a right to study painting, poetry, music, architecture, statuary, tapestry, and porcelain.

Adams wrote this in a letter in 1780. Almost 250 years later, many more of the sons and daughters of our culture are studying healthcare, arts, architecture and entertainment than study politics and war. And a good many of our young people study how to make the tools – digital and otherwise – that make the comforts that support lifestyles of ordinary people far more luxurious than past generations came close to achieving. The changing nature of work has liberated most of us from the fields to do work that not only feeds the body but extends its life, enlarges the mind and heart, and connects people to each other to make more care and meaning present day to day.

In fact, we've seen that as people spend relatively less time on physical production of food and hard goods (call these necessities), and more time thinking and studying and talking about ideas of art, science, and human purpose, we tend to create less stuff but more *tools for creating stuff*, more ideas that help us make better machines, and more schools and media and culture that reach larger and larger percentages of people. Most of which seems like genuine progress.

But like the former farmers who once understood and controlled the daily work of agriculture, and kept a human eye on things like how we treat our animals and what things we do and do not pour into the lakes and rivers running along our farms, we're already losing our collective human connection with the ethical boundaries of the work; we're already giving it over to computer systems and the machines these systems operate. When Peter drove a taxi for a living for a brief period in New York City, there were things he did and didn't do because of his ethical standards – choosing to drive at a safe speed, driving faster to get someone to a hospital, or adding an extra fare to my day's work because rain started coming down hard and it just seemed wrong to let that lady stand there waiting for the next driver on a dark and wet street. Our systems and machines have to make these decisions now, and we have to tell them how – a task that seems much, much simpler than it turns out to be.

How do we wish to live in the new age?

Consider the following example of a young professional living in a typical American city today, working at a high-end job – a job very much about ideas and communication, but not clearly connected to the ethics layers, or lacking ethic layers, in the stacks of the systems it relies on.

A young woman wakes up in her bedroom and reaches for the phone resting on the small table beside her bedside. She scans the news, then sees a note from her doctor: please call. With some concern, she sits up and calls the doctor's office. She learns the results of a recent scan of her lungs: nothing but clear and healthy tissue, a relief.

She opens her daily health app on the phone to check how many steps she'd taken the day before, and how many calories she'd consumed. She likes what she sees on both counts, and buoyed by the good news from her doctor she bounds over to the bathroom, sings in the shower, and gets ready to drive to her office.

Once there, she fires up her laptop, emails her boss and checks on a project she's leading from her office in Chicago in close coordination with a team in India. She pauses a moment to open her keystroke analysis program – a program that compares the speed and accuracy of her typing this morning to her normal patterns, ready to alert her to signals of a health concern, from lingering effects of a drink too many the night before to an undetected stroke. This program, too, gives her a clean bill of health.

Not only does she enjoy the feeling of good health for its own sake, but she's been talking with her partner about their hope to have children and they've just decided that the time is right. With that in mind, she browses over to her company's employee manual to triple-check the maternity-leave policy.

She puts in a satisfying morning's work, suits up for a lunch-time jog along the lakefront and reflects on the beautiful sun over Lake Michigan on a clear, cool day. A sharp, calming blue sits uniformly across the sky.

We can see that she lives a technology-heavy life. From the internet-equipped phone at her bedside to her medical scans and easy communication with India, she's benefitting greatly from digital tools that she's come to think of as part of the way things *just are.*

And clearly she trusts these technologies. In fact, even as she jogs around the lakeside park, she's relying on digital systems that control water levels in Lake Michigan, that keep the airplanes flying overhead on a steady and safe path to O'Hare airport, and even monitor the water from the lake that will fill her glass when she stops in her kitchen for a quick hydration break later that evening. And at every turn these systems – and many more – make decisions on her behalf that directly affect her safety, her health, her ability to connect with friends and family, and her privacy in profound ways.

Most of this decision-making is based on built-in (we could say "hard coded") choice-making logic and principles fundamental to the computer programs running all these systems. But every once in a while, that choice-making logic fails. Something that the designers or the operators of an important technology-driven system did not anticipate pops up and at the critical moment there's a failure to protect the user, or to protect society, because that specific case had not been planned for, and because rules for comprehending unexpected situations and matching them to higher-level principles, like "always charge the lowest possible price," or "protect the safety of humans in this system as a first priority," were not adequately built in.

Perhaps those basic principles, or the rules for matching novel situations to the principles, didn't make the right match. A group of informed people looking at the specific case would recognize that immediately – the system allowed levels of certain bacteria in the lake water get too high, let's say, because the system mistakenly made a trade-off between letting the water level get too high and the concentrated bacteria count without seeing an alternative that could have kept both the level and the quality of the water within their targets. Or perhaps or the higher-level principles turned out to be poorly thought-through, or more complicated than the software planners and coders had anticipated.

If the highest-level principle in the lake-water management system is to keep people safe, that principle has to be very carefully expressed to anticipate cases in which multiple threats to human safety pop up at the same time – rising water levels, threatening bacteria, too much boat traffic and nasty animal life unexpectedly thriving along the lake's marsh areas, all at once.

In these cases, we have a failure not of technology but a failure of the putting in place of human values and human purpose by the people who built the system. Values and purposes are big ideas, like "all human life is equally precious" and "preserving the quality of the natural environment for the long term is vital for our collective future." But system architects and programmers need to operate with a great deal of nuance to ensure – to code in – the right degrees of relative importance of all the variations and complications of lived experience that will challenge broad principles and demand highly specific ranking of this good versus that good, and this pretty bad outcome versus that worse one.

The task at hand is to code in broad purpose *and* lots of nuance. This task is beginning to crop up more and more in the technology-driven world evolving all around us. Its frequency and vital importance are set to accelerate enormously.

The Woman in Chicago wakes up

She looks, again, at the news on her phone as she wakes up. She scrolls through a popular social networking site. Its news-section filter automatically selects and features stories based on her profile. Is she seeing a fair selection of what's happening in her world? The news that gets served to her is intentionally slanted toward her hometown of Chicago and her interests as a young female executive who travels a lot for business – so she'll see information about weather, traffic and high-profile events in town, along with Midwestern politics, news about airports and any travel slow-downs, a bit of women's health news and high-level headlines related to business in Mumbai. But is she seeing everything? Of course not. Everything would be far too much for her to absorb. So she needs some help and the application that drives the news features of the social networking site filters the news based on a great deal of data about who this woman is, how and where she lives her life, and who her friends are.

But that only raises more questions: Does the algorithm that selects stories for her put equal weight on news from conservative and liberal politics? And how about that big space of not-so-liberal-but-not-exactly-conservative news and opinion? And how aware is she of the filtering that goes on – of how the news that she thinks of as *just there* on her phone every day gets picked from thousands of other stories, or doesn't.

Consider what will happen in coming months and years as the applications on her smart phone get smarter and smarter. The news she's reading and the message from her doctor and the information about her health and exercise habits are all running on different apps – today. But chances are pretty good that not too long from now she'll have a master application that organizes the information from all the apps and not only presents them in a consolidated form, but also shapes what she reads, controls a portion of what she knows and, to some degree, prompts what she *does*.

She might choose to have lunch delivered to her desk on Mondays – and rely on an artificial-intelligence-driven program to blend her health data, including her steps taken, her blood pressure and her pulse on a given Monday, the open slot in her meeting schedule, the current state of freshness of various favorite ingredients in the pantry at her three or four favorite lunch-serving places, and the relative wait time at each. Just as she's picking her head up from a long-distance video meeting and beginning to think about what she wants for lunch, a box arrives – the perfect meal to fit her day.

She herself will be able to establish the rules about how information from one app – like her medical records or her exercise or her food intake – should interact with the information and function of another app. That might limit or expand the information she'll be served up in her news feed, or for lunch. She'll be able to control – to a degree – the biases that she prefers in the sources of that news and the spices she prefers in her sushi. But there's simply so much news, so many news sources, such an almost unlimited combination of all the *categories* of news, *styles* of news, and *content* of news – and also *so much* sushi – that she herself will not be able to hold all of these details in mind to think through which combination, which formula blending all the variables, really works best for her. The apps on her phone, taking data from sensors in her car, her desk chair, her toothbrush and her eyeglasses will know far more about the state of her body than she herself possibly could, and they'll feed information into a larger digital system that can make choices on her behalf, based on more information *about her* than she herself can comprehend.

In fact, this is exactly the kind of large-scale data sorting that computers were invented for and there's more than enough computing power on her phone and the servers it connects to for a level of information-scanning and matching far beyond what the greatest human genius could accomplish on her best day. That is to say, our digital systems are very good at collecting, and making decisions based on, huge amounts of data and with great speed. Much better at this than we are. But knowing what's right or wrong about a news story or a sandwich is harder, whether it's the personal reaction this young lady might have to a news item about Ireland because it tells a rags-to-riches story about a wealthy business man without mentioning the brutal political violence in his past, or her reaction to California grapes in her Chicago fruit salad based on a deep personal connection to the Cesar Chavez-led grape boycotts of the 1970's. That's a bigger challenge for systems to tune into. And so, both general ethical principles and personal tuning of the ethical edge of these systems will be vital.

These are not one-and-done exercises. We don't get them right and then move on. They require constant thought and engagement with our systems, amending and course-correcting the systems as they operate, both at the ethics layer of the systems stack and at the front end, the part of the digital system that this woman sees and engages with directly. After a long conversation with a friend or teacher, this young woman in Chicago can engage the front end of her personal digital assistant to set her food preferences to "no grapes," along with an ethical flag explaining why, which she can choose to make visible to others. (This visibility allows for a social connecting and learning function important to the practice of ethics – more about that later).

So our friend in Chicago will most likely have an app in the near future for scanning the world's news, for scanning all her personal information including all her social media posts, for scanning her friends' social media posts, for adding her exercise and food habits, her health records and that daily key-stroke analysis measuring her cognitive functions up to the very moment, in order to create ever-evolving formulas for finding just the right blend of news and information flowing to her phone. A system that will work across millions or even billions of people's smart phones and kitchen computers and wrist-watch computers will be shaping – shall we say slanting? – what people know about the world beyond the small space they occupy and the relatively few events they witness themselves. An automated system. A system that starts with a purpose programmed into it, and a set of values to help it make trade-offs about what the right thing to do is when faced with tough choices. A system that operates without the need for human intervention, and without the possibility of human monitoring remotely matched to the scale of millions and billions of lives the system serves and perhaps alters because of the work it will be built to do.

A final wrinkle in this example: as this system learns more and more about each of the millions or billions of people who use it, it will almost certainly generate rules for generating the rules of which information best suits which user. That is, the algorithms that make the tradeoffs of doing this versus that will themselves evolve. This evolution of the rules will be less and less the work of people, and more and more the work of digital systems themselves, pulling lessons from truly massive amounts of data too complex for humans to make sense of.

So we will likely be living in a world shaped by algorithms that the systems themselves generate and alter. These systems start off getting their instructions from the people who write the computer code that makes them work, but eventually take more and more direction from the lessons that the systems themselves see in the vast amounts of data these systems encounter as they do their work.

This is the era of the self-governing algorithms. And this book exists in part to challenge that notion – to remind its readers of the high duty to be sure that we write our software so that algorithms are always governed first and last by human purpose and human values.

Personalized technology and its risks

When she's out jogging along the river, our friend in Chicago will be fed news and information based on what a set of decision-making tools on her phone knows about her. She'll have some ability to adjust and set parameters for how that set of formula-making tools should be working, but like most of us she will be on the other side of filters that she mostly does not control and will often be unaware of. She'll get suggestions for the right path to take her through the smallest of microclimates and mini-geographies that map her emotional and physical state – whether it's a better day for a more ambitious uphill run, or a day she'll be more susceptible to distraction, or even to the health hazards of running as part of a pack of other lunchtime athletes. Her path will be shaped by knowledge too vast for her to manage, but which her computer agent manages for her. And if she lets that system be a bit more bossy than gently helpful, instead of a kind suggestion she might find that the phone sets off an alarm when she's headed off the recommended course – or that a friend gets a notice when she pushes her heart rate too high for too long, or discover any one of a wide array of interventions her systems might make in her life.

Beyond this one young professional's jogging path, the stakes are much higher. The chemistry of the system-treated water from the lake that she drinks and the level of the of water in that lake and even the flow of traffic as she starts to cross a street will be managed by interlocking systems – and these systems will know more than any person can know. And so the limits and the rules we set for these systems will be of the highest importance.

While our first thoughts about creating these limits and rules for systems usually lead us to think about specific cases and specific rules for those cases – when the temperature is over 80 degrees, make sure you suggest a flatter, cooler route for my jog, as an example – with a bit of experience in this kind of work most people come to realize that proceeding from the example to formulate the rules in order to set limits for our systems is an impossible task. Too many potential situations keep emerging as we consider how many possible ways the forces at play in our world, from weather to microbes to political events, interact. Soon enough we must arrive not at rules for each occasion but at clarity in the values and purposes that we hold – sometimes without much reflection or thought about those values and purposes – and which people have been applying in our non-digital human experiences more or less forever.

The false idea of the self-governing algorithm

When the professional woman in our example gets a message to call her doctor's office, is that message entirely private? Let's suppose that she's an occasional gmail user – and gmail, like other free email providers, actually reads the content of all her email to help its advertisers and its sister company, Google, target advertising and service offerings. That means that who she's emailing with and what she writes are known to at least some degree by the company that hosts her email, and indirectly by all the companies that buy advertising through that company. If she's emailing with her doctor's office about an illness or about a personal health event that she considers private, can she be sure that it truly *is* private?

That will, ultimately, depend on the polices of her free email host – how fully the operators of that service protect what they know about her, whom they're willing to sell that information to, and how they allow their advertising clients to target their ads. That email host needs to know the limits of privacy law, and more importantly they need to know limits that derive from their own values and principles about how to protect the people they serve.

When she jogs on the lakefront, our friend enjoys the fruits of dozens or even hundreds of systems that help to keep the streets safe, the skies an orderly stage for carefully managed air travel, the water relatively clean, and the level of the lake just high enough to support commerce and recreation. She is also increasingly likely over time to enjoy the relative safety delivered by law-enforcement systems that help target crime, prevent social connections among convicted felons, and deploy police where they can be most effective in preventing and responding to crime. On the other hand, she might also become a victim of these systems, depending on her family's background, her personal history, and the habits and profiles of her friends. Or some combination of both.

Generally, the water this woman drinks is safe partly because of systems that monitor the level of bacteria and other key substances that could, left unchecked, create illness across the state of Illinois. But there are times when the trade-offs built into the process require clearly marked lines of right and wrong – or at least righter versus less right, or a little wrong versus a lot wrong.

Limiting the water available from the lake for agriculture doesn't seem to in anyone's interest – except that there are times when keeping the lake level high enough and the volume of water adequate to dilute potentially dangerous pollutants, water needed and expected for crops and animals might not *just be there*. If, that is, the systems making minute-by-minute adjustments are programmed to make the safety of people who drink the water a higher priority than the productivity of the farms and fields.

The question of whether this woman benefits from predictive policing programs or suffers from them is also based on fundamental values. Pioneer of British law William Blackstone wrote, "better ten guilty persons escape, than that one innocent suffer." That's a practical application of a foundational principle that can serve as a correction to the kind of tough-on-crime law enforcement that always has its advocates, if it is followed. Better to keep a woman like this safe at home by being tough on crime, because we presume, she's not a criminal. But others might say, better to be tough on a woman like this, so that the rest of us can feel safer knowing that the threat she presents is being neutralized by our police. As policing is shaped more and more by digital systems and intelligent algorithms, and police cars are assigned to monitor and protect – or monitor and intimidate – neighborhoods, streets, and individuals based on what our systems know about them; and as we lengthen some prison sentences and shorten others based on what we know about an individual's statistical likelihood to follow the law once released, we need the ethics layer in the criminal-justice technology stack to be as crisp and the application to be as principled as Blackstone's.

In this need for crisp and strong principles, policing a city and managing public waters share common imperatives. Someone must build fundamental values into the systems we rely on, and make those values clear enough that unexpected situations generating hard choices don't cross boundaries counter to the preservation of individual human rights, or the long-term importance of environmental integrity.

The work of getting these principles right – making sure they accord with the values we generally accept as a society as well as leading to the most desirable practical outcomes possible – requires two generally avoided tasks in the writing of code, designing of algorithms, and architecting of digital systems: arriving at and clearly expressing those commonly held principles across a large swath of the population, and then arriving at the fine-tuned versions of these principles that fit the tasks that systems are built to accomplish, balancing full commitment to the principle against the most practically rewarding outcome from the system's work.

The planned child in Chicago

Recall that the young woman in Chicago is planning on having a child. She's also been competing in recent weeks for an exciting posting with her company in Paris, likely a three- or four-year stint. And she has a sneaking suspicion that if her bosses thought of her less as a hard-driving executive and more as a mother-in-waiting, she'd lose out to one of the men competing for the spot. And so her online visit to her company's employee handbook – in particular the section on maternity leave – is something she assumes will remain private from her supervisors at work.

Will it? That depends on the values about privacy built into the software that helps company executives learn from the online habits of its workers – a common interest of corporate managers supported by dozens of software tools that generate reports by the week, by the day and by the hour about who visits which online resources and how much time they spend looking at what.

With this challenge in mind, the makers of these workplace analytics systems have to set limits in place based on values – they have to build ethics layers into their products – and they don't have the luxury of simply polling stakeholders and hearing one consistent answer on in-the-workplace privacy for paid employees. They can likely come up with a map of different beliefs help by different people, some clearly tied to the different interests employees have compared to their employers, but others based on personal beliefs and experience, religious feelings and commitments, and broader or narrower interest in the follow-on effects of seemingly small policies like these.

An equal measure of dissatisfaction

The makers of these systems have two tasks: first, figure out what they themselves, and the rest of us, really believe about this complex issue; and then second, to find the balance point at which they are in accord with these beliefs to the broadest degree possible, while going against the beliefs of people with divergent values to the least degree possible. That's not easy, and tends to land digital-systems makers and operators in the same uneasy landing place that social and political leaders find quite familiar: so long as everyone hates you a little bit, and no one loves you entirely, you're probably doing your job as a leader of a diverse group of people about as well as it can be done.

What we believe matters: feeling uncomfortable is a clue

If following in the cleared space behind an ambulance on a crowded highway makes you feel uncomfortable, there is likely to be an ethical principle at work in that discomfort. As important as the *feeling* is in keeping you on an ethical track in your behavior, knowing the principle or value behind it is even more important, because getting that principle right when we build software or companies – or even our families – will help us fall back on ethical boundaries when unexpected tough choices arise and our ethical instincts fail to spark, or might even mislead us. (In the case of the ambulance on the crowded highway, the principle for most of us is that we should not benefit from the suffering from others).

Cold and efficient as digital technologies may seem at times, they are compellingly human. We build them. We write the lines of code that, in their millions, make them work. We establish their limits, we shape their color and sound and 'feel.' Ultimately, we tell our technologies what to do, and what not to do. And there lies one of the greatest challenges of the digital era: not merely understanding what makes good technology, but what makes things good, more generally. To tell our technologies what to do, we need to understand how we wish to live

When we have to make our values clear in order to code them into our technologies, we will often be forced to reflect on the operant values we use but don't often think about; we will be forced to pay attention to the beliefs that have been shaping our values for a long time without us paying attention to them, and we'll have to make choices about continuing in ethical habits we have thought too little about.

Here's a quick example:

Years ago, a horrific crime in New York City had been committed in a public park that hundreds of thousands of people, rich and poor, black and white, visited every day. A white woman working for a downtown finance firm had been beaten and raped. A journalist friend of mine, on hearing the story, immediately said "that's news." Why? "Trust me – I'm a journalist – that's what people want to read about, that story makes a difference. It's news."

At the time, there were as many as seven rapes a day reported in the city – a disproportionate number of them inflicted on poor women who were not white. Why was this story more "news" that others like it? Put another way, what was the principle that drove the reporter's sense that something becomes news?

At first, the reporter said it wasn't a question of race or economic class, but just a sense that this story fit a pattern that would make the readers (and, therefore, the advertisers) of a big-city newspaper care. Thinking through the question more deeply, though, the journalist came to see that in fact that "gut feeling" and professional instinct was based on a principle that was inconsistent with the journalist's stated values of equality and concern for all people. Below the surface was a hidden value – the notion that white people and people in the higher economic classes were of more concern than others, that their stories of suffering were more important to readers and advertisers and therefore were more newsworthy, especially when a story involved the kind of lurking danger in a public park that many people worried about. In this case, taking time to think deeply and talk uncomfortably about the values below the surface of our professional habits and gut instincts revealed something ugly, and surprising.

This kind of thinking and talking can help to change a low-tech practice like deciding what constitutes "news," and it's exactly the kind of thinking and talking we need to do a lot of as part of digital ethics.

The rise of digital technologies in a sense forces us to make the effort to be better – to accelerate the negotiations among different interests and different traditions that come together in any civilization, and to bring to the surface tensions that more easily remain hidden when we communicate at lower speeds, when neighbors and fellow citizens are less visible to us and when shared value in a society rests on the kinds of work and achievements that draw on narrow technologies – like the building of bridges, the running of trains, and the support of agriculture and manufacturing. As the great works of our age shift toward digital creations which in turn become creative agents themselves, and our sense of culture and shared space enlarges to include the virtual world, forging common culture become a faster-paced and more urgent project. In the earliest phase of that acceleration, as these new digital machines enter our lives and civic imaginations, it's easy to overlook the need for a crisp ethics layer in every technology stack, but we're turning the page to the next chapter now.

Code: the noun and the verb

"Code" is both a noun and a verb in the software world. The lines of computer instructions that make up the working engines of applications like word processors, or user interfaces like websites, are code – that's the noun. But making code – writing those lines of code – is "coding," an activity. "Code that" means "write lines of instructional code in the program to make software do *that*." We "code in" the look and feel of a user interface. We "code in" faster problem solving and more elegant ways to make software do what it does.

And, just as we have to code in all these features, we literally have to code our values in too. That means engaging in a process of reflection, refinement and change.

Take again the example of the ambulance that whips by on a crowded highway. Do you change lanes to draft in its wake, filling the empty space behind it? Let's assume you agree that that just feels wrong – not quite as wrong as pushing a smaller vehicle onto the shoulder to make more passing space for yourself, but wrong enough to be ruled out. We can now code into our driverless and driver-assist systems a boundary: when an emergency vehicle passes, do not fill the space directly behind it. That's a good, clear rule. It's also a rule based on a principle: when given a choice, do not benefit from the suffering of others. That principle can apply to lots of instances beyond the passing-ambulance scenario, including many situations we can't imagine today. The principle is especially important because it will make our systems better carriers and shapers of our cultural and social beliefs, and make our systems much less likely to respond to novel situations with actions that we find threatening, unwise, or simply wrong.

But let's add some actionable nuance to this rule. Is that an "always rule," like "never kill a child?" Indeed, is "never kill a child" itself an always rule? It seems more clear-cut than just "never kill" because most of us can imagine a circumstance when killing would be a terrible thing, but not as terrible as letting a man about to kill a dozen or a hundred others go about his business when we can stop him with lethal force, with just a split second to decide.

Make the man a child and most of us will hesitate. But reflect further – the child about to kill a hundred or a thousand other children – and we might choose to put that rule in the "sometimes rule" category rather than the "always rule." In fact, very few rules seem upon closer inspection to fit into the category of "always rules," and perhaps no rules might belong there.

So not only do we have to clarify in this case of the ambulance speeding past what the underlying moral principle actually is, we have to decide the strength of the rule that comes from that principle. "Do not benefit from the suffering of others" is an attractive principle, but it might be best categorized as a kind of "sometimes rule" that hinges on the relative cost of following that rule. Perhaps we want a formulation something like this: Do not benefit from the suffering of others unless the benefit is great, and also accrues to others.

Here's the thinking behind this nuance. This moral principle about not benefiting from the suffering of others is clearly about putting the wellbeing of others ahead of one's own wellbeing. If we feel no moral restriction or discomfort about benefitting from the suffering of others, we'll allow and at times create more suffering because in perceived zero-sum situations other people's loss makes my gain more likely.

By following a moral principle of not benefitting from the suffering of others, we reduce the net amount of suffering of all people, or at least make the net reduction more likely, especially if this principle is a norm that large numbers of people follow. It's a distillation of a communal concern and regard for others: I don't want to benefit from your suffering, and I don't want you to benefit from my suffering.

Perhaps I'll try to make this principle such a pervasive norm in my family and among friends that we'll all be more likely to absorb it as a habit or "instinct" – one of the principles that most of us might not even know we have inside of ourselves until that ambulance whips by and I think about jumping into that open lane but I don't because it just doesn't feel like the right thing to do. That feeling of right and wrong is the effective function of that moral principle, once that principle has become pervasive in our lives and our micro-societies. Perhaps I learned that principle at my parents' dinner table, or over months and years of following their example. Perhaps I learned it at church or temple or mosque. Perhaps I reflected on it because of a philosophy class, or exposure to a friend with a strong conscience and a taste for these kinds of discussions. However it got there, there it is. It acts on me in that moment of sudden decision. I choose not to swerve into the open lane.

Shall that principle always apply? Especially because it is a principle about not benefitting from the suffering of others, in this case the "sometimes rule" effect might be measured in terms of the benefits to others too. If I'm delivering an organ to be transplanted to an otherwise terminal patient waiting at the end of my journey, I might immediately feel that it's OK to leap into that open lane. Upon reflection I realize I've followed a limit to the sometimes rule: if someone other than me benefits greatly, don't follow that rule about the not benefitting from the suffering of others. Strong good for others comes from this exception, and reflective thought in advance has made clear that this is a rule that works more in the abstract and general (unlike a very specific rule, "Don't ever hit your sister," for example). But this exception itself must be limited: let's not run over a group of children to get that organ to the patient faster.

An important step in the actual work of crafting the ethics layer to a digital system – after a principle has been clearly identified as important to include – is deciding on how strong the principle should be: one with no exceptions (an Always Rule)? One with some exceptions (a Usually Rule)? One with many exceptions (a Sometimes Rule)? And then, logically, we need to clarify and express what the logic needs to be for identifying exceptions. The exceptions logic, like the principles themselves, may be revealed by working through specific instances, but should be as close to principles themselves as possible. The goal is not to code-in a catalog of principles and exceptions, but to code-in the kind of principles and exceptions that are generative, that can apply, ideally, in all cases without listing those cases in advance.

The ambulance-lane example is a good one. The general principle in this case, to not benefit from the suffering of others, is probably a principle that we'd want to have more exceptions and less overall weight than, say, the don't-kill-children principle. Not benefitting from the suffering of others seems like a Sometimes Rule. Consider program-trading systems that make money by going short as well as long in stocks, and businesses in competitive markets generally. When my competitor closes her shop down the street, we can presume that she and her backers suffer in some way. Should that stop me from trying to win this street-level battle between competing shops? I'd think not. So we look at fair and legal business competition as an exception to the rule here – and in fact as a whole category of exceptions. What's the underlying proposition to this exception? A strong argument can be made that the larger economic system is made stronger and the large body of customers better served though fair competition. The greater long-term good for a great many here drives the exception.

The business notion of "thought leadership" is important to this conversation – some individuals will become explainers and champions of ideas they believe in. Others will put their efforts into documenting who really holds what opinions about key issues across large populations. Government leaders, generally with an eye on pleasing the largest numbers of potential voters, will on rare occasion take the risk of trying to codify these beliefs in law. And makers of systems will often tweak the ethics layers in their wares to reflect the beliefs of their customers.

Ideally, these players on the digital-ethics stage will have strong, positive influence in taking the digital ethics project forward. Dangers lurk around every corner, though. When political leaders feel they can gain more by pleasing big-money players than by engaging and bringing together large groups of ordinary people; when sellers of systems gain more by selling *access to* consumers rather that delivering value to them, we must see these dangers for what they are, and must seek to balance them by the most democratic, broadly inclusive, loudly public conversations we can have.

Finding the edges of our beliefs

Particularly in the case of Sometimes Rules we need to use the philosophy tool-kit to discover the edges of our starting-point beliefs through extreme examples. Think of cases that bend or break the principles that feel generally right, and do the hard thinking work to understand why this or that case should be exceptions – to understand what it is about the exception that might apply to other times and places that should be exceptions too. The trick is not to describe the one exception, but to capture the spirit of that exception and the good it speaks to.

The issue of gun safety can illustrate the work of finding the right ideas and language to express important exceptions to important principles. While gun safety might not seem like much of a digital-ethics issue today it is already the focus of an important debate made particularly visible by Harvard Law professor Jonathan Zittrain. Noting the shockingly high proportion of guns stolen for illegal purposes – from a thief stealing a registered handgun from its owner's home and then using it in a killing, to paying off a government official to mis-route long guns meant to arm a police force – Zittrain, whose work sits at the intersection of law, technology and ethics, observes that gun manufacturers could be required to build in biometric sensors that would allow only registered owners to fire their guns.

When a gun is sold, a record of ownership can be collected and controlled by a government agency or public-interest registrar. From that point forward, only the authorized owner can be allowed to trigger the digital safety mechanism and enable the gun to fire. Members of a police force or military unit could have special status to link them to groups of weapons assigned to their units. The technology is entirely plausible based on digital systems today, though it would take some work to establish standards and institutions to put it into force – and of course it would require a political will absent as of this writing in the political leadership of the U.S. and other nations as well.

Digital technology could plausibly make it almost impossible for unregistered users to use guns to kill or to harm others. It's also plausible for sensors in guns to make suicide much harder, by matching biomarkers of gun owners with proximity sensors. No only could these systems prevent a gun from firing in anyone else's hands; they might prevent the gun's use in suicide – the majority cause of death from firearms in the U.S. today.

What would the ethics layer of the gun's digital-safety system need to include? If we begin with a bedrock principle like "all human life is equally precious," we quickly have to confront the fact that the very purpose of the gun presumes exceptions to that rule: the person firing is potentially ending the life of another person. That shooter is putting his life or someone else's above the life of the person at the end of the gun barrel. So "All human life is equally precious" is a *sometimes* rule, but a strong sometimes rule with few exceptions, if we intend to have working firearms in our world.

Whom do we trust?

As we consider these exceptions, we realize quickly that the core question is trust in the gun user as judge of who might or might not get shot. Not just any shooter is entrusted, but specifically a *designated shooter*, for whom the digital system effectively unlocks the gun. In the case of a police officer intervening in a potentially deadly crime, an exception based on trust might look like this: All human life is equally precious, *except for when an officer of the law feels that potentially ending one person's life will save someone else's life or accomplish a good of similar value.* We trust the officer's judgement in a heated moment that the life being saved warrants saving, and the life potentially being ended is in play because of some threatening action that individual has taken.

The systems architects and coders here are likely to presume the ethical standing law-enforcement agencies generally. Unless the digital system includes an ethics layer that digs more deeply into how and why law-enforcement officers choose between good guy and bad guy, the digital system affirms and reinforces both the good and the bad of the existing criminal justice system.

Doing the deeper ethics work

Should digital ethics as a practice avoid deeper engagement with past challenges and inequities? We can, of course, take the position that we want digital-ethics work to do relatively little but to do it very well, and any work of clarifying and codifying ethics gets riskier and riskier as it grows in scope. Nevertheless, when we have the chance to revisit the way things have been and make them better, we should. That might be the highest principle of all, the most fundamental of all the layers in the stack.

The goal is not to make a bridge between some abstract desire for social justice and the specifics of what a digital system does. The goal is to have the principles that we work hard to clarify and express – the principles that we believe in as a civilization – honored not only in terms of what we and our systems do next, but in terms of what we allow to continue that might have begun long ago.

A broader principle allows a digital system to support a police officer using the gun under specific circumstances but warning her off under other circumstances. We might craft that broader principle to say something like "gun users should only fire their guns at people when firing the gun serves a planned and legitimate purpose." This has the potential to be an always rule. It might mean that the homeowner who is licensed to own and keep a gun specifically for home protection can fire the gun within a short distance of the home, but not a mile away. A similar restriction might bear on a business owner or a security guard. And even the police officer would face restrictions – and find benefits – from an always rule like this.

In many cases, police fire at people they perceive to be threats because they judge in a fraction of a second that what looks like a gun in a person's hand *is* a gun, when it might in fact be a toy, or a flashlight, or a wallet with a card reading "I am deaf" (a classic police-training scenario). An always-on digital network identifying the location and status of firearms can be a vital decision-tool, based on that principle – gun users should only fire their guns at people when firing the gun serves a planned and legitimate purpose. Of course, this technology and these guiding principles do add new kinds of risk – like system accuracy – to the already risky job of the police officer.

With all this in mind, we want to affirm again that when we have the opportunity to make things better, we should.

Chapter Four

Tuning In   Some Real-World Examples of Learning to See the Ethics Layer in the Stack

Three presents for Peter's children

When my children were young, friends would sometimes bring them gifts, always with a wonderful spirit of generosity. Our oldest daughter was once given a small model of a ballet dancer, cast in metal – maybe bronze. It was beautiful, and an impressive piece of technology. Clearly not carved from wood, how was it made? She could see it on the shelf and wonder.

By the time the children were about 10, 7, and just shy of 1, someone brought them a fascinating hard-plastic ball with rubbery bumps and projections popping out of its surface, overall about the size of a cantaloupe. Somewhere tucked between these odd protrusions on the surface was a tiny switch. Turn the switch and the ball began to shake almost violently. Place it on the floor and it would lurch about making lots of noise and pulling and tugging this way and that with seemingly manic energy. Just holding it was electrifying, and the kids grabbed it from each other utterly transformed by the bundle of wild energy they would try to hold and hug (and, in the baby son's case, try to eat).

A third present was a small, fast-spinning turntable with a spindle point in the middle. Spike a piece of paper on the spindle, press it flush against the surface and the paper would spin. The toy came equipped with small squirt-bottles of paint. Hold the squirter over the paper and squeeze: an eccentric and often dazzling little painting would emerge.

Each of these presents not only had a distinct kind of magic for these three lucky kids, and each was very clearly *for* something different from the others.

Of course, what a toy is for can mean two different things: what it is intended to be used to do, and what it might advocate in some sense.[1] What's this paintbrush for? It's for spreading

---

[1] I owe special gratitude to my brilliant friend and colleague Karen Morris for her launching of the conversation about the two sides of

paint on paper, or on the wall. But what's this beautiful little statue for? That's a harder question because it's not a practical object. It's for looking at. But it does suggest some values, too. In a way, it's an argument in the form of an illustration, advocating beauty – the creation of it, the preservation of it, the importance of having beauty close at hand even if small in scale.

The object you hold in your hand is, like software, the result of a stack of values that begins with a first layer that declares how important beauty can be even in the simplest object that does not ring or sing or entertain in any way beyond being, simply, beautiful.

Similarly, that spinning turntable was for the creating of little paintings and having lots of fun in the process. But it's also a hands-on, you-create-it kind of toy luring kids to become creators and artists even if they think they can't or don't want to. That's a value and an act of advocacy as well. It's for making little paintings, and it's for getting kids to be active rather than passive in their play. The stack it is built on begins with the primary value that kids should make things as the center of how they play.

It's hard to figure out what that crazy shaking blue ball is for, in either sense. It's a good example of the kind of technology that makes you say *wow*. It's lots of fun to engage with but doesn't solve a problem, and it doesn't engage or connect people with one another (other than, in Peter's kids' case at least, to

---

the question "what is this for?" particularly in terms of what companies and other large organizations.

want to grab it from someone else and hold it tight). It just kind of sits there and shakes in a weirdly attractive way.

If you're a parent, thinking about what toys and games are for is an important exercise. Peter and his wife had fun with the blue ball (which his kids took to calling "Wild Boy"), but valued the little statue of the dancer more, and absolutely loved the turntable best of all – because of what it was for in the larger sense, what values it brought forward. They wanted their kids to be active rather than passive, to be creators and artists rather than competitors for a shaky ball that could offer a selfish thrill.

What's a smartphone for?

So "what's this for," in the larger sense, is a vital question. And that extends well beyond children's toys, though as users of newly emerging technologies we all have a lot in common with new parents – we're doing many things for the first time, trusting some tools and technologies we haven't used in the past, and figuring things out as we go.

Consider, for example, how a seven-year-old child might use a smartphone. What's that phone for? Well, it's for making phone calls, going online, texting, shopping, calling for help if needed, being locatable in an emergency, and lots of other things. What's it for, in the other sense? What's it in favor of? It's for keeping friends and family connected – that's a value as much as a function. It's for empowering a child and making him safer by keeping him connected in real time with a family and a society that cares about his safety. And it's for encouraging her to look up answers to questions about how the world works, in real time, when facing new interests and unknowns – there's a positive value about learning built into that device.

Both sides of the "what is it for" question connect: the values we build in to the technology shape the ways that we use the technology – and can do so with even more effect if we emphasize those values. If we talk with our kids when we give them smartphones and explain that while its uses are limitless, it's there to keep us connected and to help the keep possibility of actively learning present all the time, then we know it's OK to make a rule that says no phone time when the family's together talking, unless it's to look up something we're talking about. We use the values dimension of "what's it for" as a guide to shaping and explaining the functional side of that question.

Does "ethical" mean good?

We often make the mistake of believe that being "ethical" is always good – that the problem with bad actions is that they are without ethics. In fact, being ethical means following a set of principles and expectations. Sometimes, if those base values or first layers of ethics stacks are not truly what we feel is right and good, the whole stack will become a machine of sorts to create bad outcomes.

Consider: The Bay of Bengal in 1943

The Bay of Bengal is an unusually beautiful area, touching on four countries – India, Bangladesh, Indonesia, and Sri Lanka. An enormous stretch of shallow ocean runs out from the sandy beaches along much of its Indian coast. If you have a bit money in your pocket, you might enjoy a few nights in one of the quiet resorts in Mandarmani, on the coast about 100 miles south of the old city of Kolkata.

The rural area stretching out below Kolkata is marked by the rhythms of the seaside, near the border with Bangladesh – a cultural and geographic calm broken every year or so by wicked monsoon rains and every century or so by violent religious and political upheaval.

The urban core of Kolkata has a more striking history. The image of the city long known as Calcutta in the West has been as a place of privation, starvation, disease and government neglect. The *New York Times* ran a series of articles about homelessness, crime, and a general feeling of social decline in New York in the late 1980's, under the banner "The New Calcutta," perhaps unfair to New York but certainly a thoughtless slap at the Indian city.

In 1943, the entire region of Bengal – from the beaches south of Calcutta to the hundreds of miles of thinly inhabited villages stretching north toward China – suffered an extreme moment of privation that reinforced its reputation. Famine swept through the province, killing three million people out of a total population of 60 million. Virtually every family, every town and every neighborhood was touched.

World War Two was part of the story. England was still the colonial occupier of India, significant fighting was underway in nearby Burma, and the Japanese occupation of China had ripple effects through Asia. Food production fell as farmers took up arms and went off to war, though the worst effects were felt by families of traders and artisans whose customers fell away as they devoted all their money to rapidly inflating food costs.

A debate continues today about whether the three million lives lost to starvation in Bengal in 1943 were the result of a "real" famine – a dramatic reduction in the amount of food available across the whole population of the area – or a failure of social and market forces to make stockpiled food available to those who needed it. Certainly the failure of a rice crop in the winter of 1942 caused by monsoon rains, cyclones and tidal waves in shoreline agricultural areas set the stage. Also in 1942, the Japanese invaded Burma and cut off the Burmese rice crop from Bengali traders. Waves of refugees from Burma added to the food demand as well. By early 1943, the famine was evident in the streets of Calcutta and the villages spreading south toward the coast.

A boy of nine witnessed the famine and eventually became its great chronicler and explainer. Later this boy, Amartya Sen, would win the Nobel Prize in Economics for related work. Sen studied the famine and determined that, in fact, there was more than enough food to feed the millions who perished in Bengal in 1943. The problem was with competing demands from local traders and clan leaders who were inclined to hoard food as war came near, and from the imperial British government valuing support of its troops – money and food that would otherwise have sustained the starving Bengalis – over the lives of its colonial subjects.

Sen's extensive research documented that today, with very few exceptions, when populations starve the failure is fundamentally political and not agricultural. He acknowledges that crop failures, extreme weather and bad planning often result in food shortages, but even when food is scarce in a region, enough good will and surplus food exists elsewhere that in almost all cases money and food are available to prevent large-scale starvation. To oversimplify a bit, in the world we live in today when populations starve what's missing is not so much the food but the will of institutions – political and economic – to help them. That's the big idea that Sen's careful documentation established.

The relative value of human lives

The question is whether governments and other social leaders value the lives of the people who are hungry more than they value the food and money they might keep for themselves.

Once we can isolate this question about valuing the building blocks of wealth more than the lives of poor people as an ethics-layer question, we can begin to see the root of the disaster in Bengal in 1943 as a direct result of a well-functioning system built on a terrible ethics layer – an ethics layer that makes the military success of the colonial power a higher good than the lives of large numbers of colonial subjects.

This matters enormously for digital ethics because in an era of breakthrough technology growth we can too easily overlook the fact that political problems require political solutions, and technology solutions to political problems just don't work. Ideas and values matter – not just technology. Thinking about social values and the effectiveness of our institutions is vital for the value of technology to emerge and actually help us live better lives. Ideas and values surround our technologies, they support them and extend them and determine whether the *value* of culture's greatest creations will reflect the *values* that set them in motion, or not. Deeply experienced and very well-intended technology leaders often hit this kind of wall when great technical innovations encounter ideas or values that we might fairly call broken.

Knowing it when we see it?

U.S. Supreme Court Justice Potter Stewart planted a flag for this kind of poor ethical decision-making when he wrote his landmark 1964 decision that while he could not define hard pornography, "I know it when I see it." Far better to craft principles that reflect our values, and apply them not only to explain what's wrong but to set out guides to our future behavior so that those principles can help us make decisions when our "guts" are confused or unreliable.

For both the store-house decision maker choosing to hoard food that could and should save lives, and for the government decision-makers who put politics and prestige above the practical interests of people they serve, we can propose a useful ethical guideline that tweaks the utilitarianism of the philosopher Jeremy Bentham. We might want to put it this way: "People with responsibilities for others need to make sure that they seek to do the greatest good for the greatest number of those they serve." If we believe this, we not only explain why we think these two leaders failed the people they served, but we can now apply it when we make decisions about new situations and new technologies.

Define Good

This finding of, and expressing, of principles we believe in so that they can become a guide for future decision-making is absolutely vital – and too little practiced.

Cow or puppy - the ethics of the unexpected

Imagine a lovely day in a quiet corner of a country known for its seaside villages and museum-filled cities. It's the kind of place that people from more competitive and money-oriented places come to when they can get a week off. On this day, the sun is shining, the air is clear and the road stretches ahead like a ribbon through the green hills overlooking the sea. There's no traffic to speak of, just a single cargo-van ferrying groceries from a warehouse to a small shop in a quiet seaside village.

It's the kind of day that might tempt the driver's mind to wander, awash in the calm landscape, but this van has no driver. It's an autonomous vehicle – a driverless car – and it won't fall asleep or pay attention to distractions or day dreams.

Suddenly a large animal darts from a low spot beside the road directly in front of the van, with maybe fifteen feet between the front bumper and the looming beast. But there's no shock or fear at play here – no slamming on breaks or wild yanking of the wheel. The car's sensors see the beast; in fact, it's a cow, startled from its own midday wandering by a snake in the grass (remarkable how quickly the otherwise slow-moving cow can move when met by an unexpected snake).

With a refined blend of urgency and calm, the van slows quickly but not too quickly, and moves from the right side of the road (head-on the to the cow) just a few feet to the left, where the sensors have already made clear no other traffic is in the way or coming soon. But the grass must be teeming with unseen drama today because just as the van is a few feet from clearing the cow, another creature leaps up to the road – a coyote? a mountain lion? No, it's a farmer's dog, not much more than a puppy though large enough to change the physics of what the moving vehicle ought to do next. There's simply not time enough or road enough to slow or swerve to miss it without going sideways into that unfortunate cow now looking more like steak dinner for half the village awaiting its groceries.

What the van does now is a matter of choice – not the choice of a driver (there is no driver) but the choice trained into the moral engine of the vehicle. That moral engine is just as important – in fact, we want to assert that it is more important – than the physical engine powering the wheels that turn to drive the van forward, because that moral engine contains the limits and some sense of the purpose of that van.

The van can do many things that the people who use it, as well as the people whose lives it touches directly or indirectly, don't want it to do. When a man or woman sits behind the wheel of a van, the sense that some things one can do with that van – drive 100 miles an hour, crash into a house, block a busy road – are wrong is, we hope, wired into the driver's mind. When there is no driver, the people whose work creates and sustains that van have to think quite explicitly about those things that we don't want the van to do and train the algorithms ahead of time, in digital patterns that shape the way the van acts and reacts in all of the circumstances it encounters.

The rise of driverless cars, just like the rise of a mountain of related digital technologies, forces all of us to step back and decide explicitly what we think is good and bad, what actions are OK and what actions have to be prevented when machines – machines like cars and machines like enormous databases – act as they've been programmed to act.

The Trolley: asking questions to reveal principles

Ethics builds from broad principles of right and wrong, adds the logic of cause and effect, and pays close attention to new challenges and new kinds of decisions that people make as the world around them changes. It's never enough, we realize, to have a list of what is good to do and what is not so good, because the world will serve up new choices that show every list, now and then, to be incomplete. Ethics must be generative: not only helpful in terms of hard choices others have faced in the past, but useful for understanding challenges new to us, and new to the world.

Just like the coding that our driverless van on the sunny seaside road requires before we launch it into the world, the ethical challenge of choosing between two or more entirely unhappy alternatives when more pleasant options are not available is a very important one. As the Harvard business-ethics professor Joseph Badaracco has insightfully said, the harder work of ethics is not choosing between right and wrong, but often the necessary choosing between wrong and wrong.

A couple of generations of philosophy students have encountered what was known as "The Trolley Problem," beginning with the work of British philosopher Phillipa Foot,

* If we make the massive assumption that people get this stuff right, how do we do it? If it is logic not magic it should be possible to replicate digitally.

laid out in a 1967 article. The Trolley problem is in fact an early version of the driverless van dilemma. It runs something like this: You find yourself walking along a trolley track and you notice that the track splits in two not far from where you are standing. The track actually takes the shape of a large Y, with the main line splitting into two right next to the spot where you're standing. In fact, a rail switch is just an arm's reach away from you – a tall wooden lever which, if pulled, will direct a train coming down the main track onto one of the two branches.

As luck would have it, you hear the ringing bell of a trolley, and see it coming in the distance, heading toward the splitting-off point beside you, where one track becomes two.

You notice that the switch is set so that the train will pass onto the track on the right side of the angle. Suddenly, though, you see a group of small children playing on the tracks – several, in fact, bunched together playing games right there on top of the right-side track. You also see that one child who has wandered over to the other branch, the left-side track, looking off into the far distance, away from the train.

You call out to them. They don't hear you. You look back at the trolley which isn't slowing down. It will be at the junction point in mere seconds, clearly moving too fast to stop in time to avoid running over the group of children (in one case), or the single child (in the other). If you do nothing, it will run over the children on the right-side track. But if you pull the switch right in front of you, it will run over the single child on the left-side track.

Let's suspend time for a moment and leave this scene temporarily — with the train bearing down, the cluster of children on one extension of the main track and a solitary child on the other, and you standing there, your hand inches from the switch that could move the train's path from the cluster of children to the single child.

Think for a bit about how a teacher might present this developing moral puzzle to a class. She might start not with the story — with moral puzzle itself — but by asking a few key questions to elicit core beliefs among the class, to establish a clear baseline of what we say we believe, even if that belief is itself fuzzy at first.

If we ask a group of college students whether, when forced to choose, it is better to kill or let die 100 people than just one, most will be quick to recognize that while both are terrible choices, it is better to lose fewer lives than more. Most will begin by saying that it's clear which is the less terrible choice, though it might take a bit of back and forth for the essential value, the first layer of the stack, to emerge: because human life is more or less of the same value, deciding this question based on quantity of lives instead of quality of lives makes sense.

We can force students to face some of the nuances by asking other questions — what if it's twenty sick, elderly people versus two healthy youngsters? What if it's a hundred MD-PhD public-health researchers racing for the cure of a spreading plague against one professional dog-walker? Other, alternative first layers can rise to the surface through this kind of conversation — for example, "Lives dedicated to service to others are more valuable than lives that leave no impact," or "People with more years likely to come in their lives deserve more consideration in a crisis than people who are likely to die sooner."

Finding some level of consensus about which of these variations is best is not the most important point of this kind of exercise, though it certainly is important. Most important is that these kinds of conversations help us clarify what we really believe, and how these beliefs create what we think of as the instinctive or "gut" reactions to questions of right and wrong. These conversations help create the habit of reflection, refinement and reinforcement of our beliefs and help us to act more in accordance with our beliefs. And they help us to clarify the values we can then code in to our systems and machines.

What if you inflict harm by doing nothing?
Part of the challenge of the Trolley problem is that it raises the prospect of failing to act when an action can help reduce serious suffering about to happen close at hand. It's easy to set up the variables in the Trolley problem so that most people will feel that doing nothing will be irresponsible – put an angry junkyard dog on one of the tracks coming out of the split and a group of infants in baby carriages on the other. The inclination to do nothing in the face of a hard decision is common. But it holds the potential of failing to act in a way that most of recognize as indirectly causing harm.

Should it make a difference whether we inflict harm by actually doing something – raising up an arm and striking someone, for example, or taking a shield out of an innocent person's hand – or by standing by and doing nothing when a small action might prevent someone's suffering? In most cases most people believe that we have a duty to act to prevent harm, but that passively letting harm happen is less wrong that actively doing harm.

Let's assume that these two questions – is all human life equally precious? Can doing nothing be bad when we have the chance to help others? – were the very questions a teacher had led a group to discuss for an hour or so prior to introducing the Trolley problem. Next, the Trolley problem is richly described to the class: the fast-moving trolley the one track branching off into two and the switch right there, directly in front of you. And the children on the tracks, just beyond the junction point, just beyond hearing your warning, oblivious to the deadly danger before them. The hand-operated switch is right there within reach.

The time is now: the train will be past you in a second or two. You can pull the rail switching the train's path, or you can do nothing, and the group of children perishes while the single lucky child on the left-hand split will live.

Doing nothing is perhaps the easiest path in the moment: the more vividly we paint this picture, the more students squirm – and more than squirm. Some turn red. Some literally become ill, imagining the enormous ethical burden of that awful blend of knowledge (you see the children), power (the switch is right in front of you, and has a clear use that will matter a great deal), and principles (you know you should act to save a life, or lives, if you easily can; and you know that all else being equal, four or five lives make a greater ethical claim than a single life).

We've been in the room when a distinguished professor of philosophy at Stanford University used his charms to paint so vivid and unavoidable a sense of ethical duty to do *something* awful that a young woman took to her feet, pointed and yelled at him with no sense of irony at all, "you're a bad man!" She surely meant it.

We need to practice

Face these dilemmas we must, difficult as that might be, if we wish to act ethically. We need practice in understanding that our principles will require action at times. And we need practice in order to recognize that when our principles conflict, paralysis because of that conflict often leads to terrible, preventable outcomes (as Joseph Badaracco has written, ethics is seldom as simple as choosing right over wrong; often it involves choosing one wrong over another).

So we are faced with the train, the switch and the children on the track. The most common outcome of this discussion that I have witnessed is the reluctant conclusion that you must pull the switch, saving four or five and dooming one. Many people presented with this ethical challenge for the first time will leave the discussion feeling shaken, having faced a vivid example of ethical duty with a high ethical cost – feeling that they now have "dirty hands," but had no choice in this example. Many find it intellectually exciting and morally important, but few actually enjoy it. Many, in fact, despise the experience. I've spoken with military and business leaders who find this example a useful one, capturing a bit of the spirit of what it is like to make difficult decisions like sending soldiers into certain harm, or firing one

* Generally we do not want to face these choices or admit that they are made on our behalf. That when we have our hip replacement, somewhere a baby dies.

group of workers in order to save the jobs of another. No one feels good about those actions. Third alternatives are almost always searched for with enormous effort and – at least at the discussion table, with the parameters controlled to force the ethical choice – not found.

Would it matter if one group on the track is young children and the other group all older people? A man on the right-hand track in excellent health, and one on track two dying of cancer? One the parent of a young child and the other childless? One rich and one poor? The key elements here do not change: knowledge, power and principle combine to demand some kind of action. That is ethics.

There is a timeless quality to this example. Trolleys and railroads have been present in much of the world for about two hundred years now, and one can imagine a not-too-different variation of this ethical challenge stretching back two thousand years, involving a run-away ox-cart perhaps.

The Hebrew Bible, in fact, offers a pointed bit of guidance to Jewish armies: when surrounding a town in time of war, don't poison the wells, and leave an avenue of escape for civilians trying to flee. These acts of humanity run against the temptation to do long-term harm of innocents, for the sake of winning a battle. The battle might seem itself to have the highest importance, but the more foundational ethic of human decency extended toward civilians even in time of battle is even more important.

Who owns the trolley?

Sometimes these ethical situations are surprising, reflecting principles accepted in one culture but not in another. For example, how many people upon hearing the Trolley problem ask, "who owns the trolley?" Not a lot; once in a while someone does. Certainly some ancient ethical texts look to the ownership of the run-away oxcart as an important piece of the puzzle: is it the King's cart? The feudal landlord's? A widow's?

Considering the history of the railroad, what would have happened to a slave in the American south who pulled a switch changing the direction of a train bound for a marketplace, on a tight schedule? What would have happened to a feudal serf in Russia doing the same? We know, as well, that in many times and places, one child's life has not always been equal to another's – social station, race, gender and a dozen other factors might lead the ordinary man or woman to apply principles to the ethical equation that we would simply not accept ourselves.

Who needs to talk about these things?

All these concerns belong on the table, and all need to be talked through to reach the kind of consensus about values that enables good code that can create good ethical layers for our digital-system stacks. The questions of how big a table we need to sit around, who should be there and who gets to decide the ground-rules of the conversation are absolutely essential. These questions are also filled with traps. It's tempting, for example, to fall into the hole of asking "Who decides who decides" as we frame these conversations. If we say, "We want as many people at the table as possible," who is the "we" we're starting off with?

Do we start by saying "Everyone decides who decides?" Does that mean we force people into the conversation in some way if they don't volunteer?

In the world of artificial intelligence, this question is more difficult still. Values and biases can already be hidden in the data that digital systems make decisions based upon. How data has been collected, excluded, and processed; where the data comes from, how it got there, and whose interests have been prioritized throughout that process are *all* vital questions. They remind us, and we must continue to remind ourselves, that there is no such thing as completely neutral data.

Chapter Five

Who Owns What in a Digital Age?
The Year 1689 Explains Why A Son is
Angry with His Father

British philosopher John Locke wrote in 1689 that "though the earth, and all inferior creatures, be common to all men, yet every man has a property in his own person: this no body has any right to but himself." Locke was laying a fundamental plank in the foundation of a new philosophy – the philosophy that Thomas Jefferson carried forward in writing the *Declaration of Independence* almost a hundred years later, and the very same philosophy that much of the world now regards as natural and proper today – the philosophy of human rights, of the dignity and integrity of the individual. Locke was decisively writing *against* slavery, *for* limits of government power and *for* the importance of the rights of the individual. Of course, he could not have known that Thomas Jefferson would champion Locke's idea that all men are endowed with the rights of life, liberty and, as Locke wrote, property (though of course Locke meant the property each man has in himself and the Southern slaveholder's notion of one person owning another).

In some ways, Locke was writing in opposition to the transgressions of the individual and the individual's rights that were all around him in England's monarchy; he was writing in his moment, about his moment.

In other ways, Locke was part of a centuries-long development of ideas. He carried forward the growing emphasis on individual rights – at least rights for *some* individuals – that arose alongside the slow-motion technological progress visible as early as England's Magna Carta in 1215, including the invention of the windmill, large-scale commercial credit and the technology literally supporting the London Bridge completed in 1209.

Locke was explaining how a good government ought to work as ocean-spanning ships and new understandings of the planets, the oceans and the heavens led his countrymen into the world's first spasm of globalization.

Perhaps Locke knew that his ideas would be part of the debate about his nation's colonies in America, where ideas about freedom and individual liberty were already being tested. But he certainly did not know that he was also writing about why Peter's son – a sharp lad of 18 – would be so angry with him roughly 300 years later as he pressed the glass surface of his phone and posted an image of the lad, looking strong and happy after a hike with his father, to Facebook. He thus published that image to about 500 of his "friends," a group sure to include not only much of his extended family but also a couple of his son's buddies as well. And from there who knows how many reposts that kind of snapshot might merit.

"I told you," he reminded his father with barely contained anger and clearly thinning patience for a dolt like Peter, "you need my permission if you're going to post a picture of me." Indeed, he had told Peter that, and Peter was right there on the edge of parental ethics, thinking, Man, this is just such a great shot and he looks so good, how could he mind, really? What harm could it possibly do?

Answer: it violates his proprietary sense of himself, just as Locke wrote. But when Locke talks about a man's right to "his own person," should we take that to mean not just his body but his image as well? We think so. Locke could have chosen the word "body" but chose a word instead that even in the 1600's meant more than a man's head, shoulders, knees and toes. One's person is more than one's corpus. It is the root of one's personality, the distinctiveness of what makes me me and you you.

Locke was writing philosophy, not law, and this issue is still far from settled. But we do need to think through a few principles to help manage questions like these: Who can post images of whom on social networks? Who can capture images of ordinary people in public places, and what can they fairly do with those images? How much of what a business – or government – learns about us in the process of giving us what we pay for, or what we vote for, or where we go in the world or online, can that business or that government use without our permission or knowledge?

The young man's Lockean impulse is not a bad starting point, and it brings us right back to the first layer in the digital ethics stack.

Stealing sound and images: How much would you pay for "Feelings"?

Great museums, of course, invest in great security. Try to lift a painting from the wall and you can expect a hand on your shoulder before you feel the weight of the frame shift to your own sticky fingers. Try to sneak in at night and you'll need the skills of a ninja and the good luck of a lottery winner to find your way back out without a pair of handcuffs on your wrists.

Imagine, along these lines, a young man focusing his talents on gaining possession of a priceless painting hanging on a museum wall, in a town large enough to host such a treasure but small enough to have good but not world's-best security. He wears a disguise – an expensive fake moustache borrowed from an actor friend, wire-rim glasses and fake mole penciled in on his cheek. He knows he'll be filmed by the museum's security cameras, but his moment of crime would be so fleeting that he fully expects to do the deed and walk right out the front entrance. Yet when the moment comes, though he stashes the fruit of his bad act in his jacket pocket less than a second after the grab, he is nabbed almost instantly. Three steps into his escape, two men in matching museum-security jackets appear in his way and ask him to please come along with them.

"No photographs" is the policy in this mid-size city's finest art museum, a policy on the way out in most of the better museums but still in force in many. These images are the source of our value, museum directors have explained. And if anyone can take the image, we won't be able to exist.

The young man is hardly innocent. His phone's camera is itself a work of art, modestly expensive, capable of capturing a remarkably fine image of a painting, suitable for reproduction. Indeed, his plan had been to reproduce that image with a partner and put out thousands of high-quality posters to be sold on the streets of his city – including a spot right in front of that very museum – for a nice profit.

Ordinarily, copyrighted images, including many paintings in museums, can only be reproduced with the permission of the owners of those copyrights, and often that permission comes with a price tag. Until recently, even an older painting not under copyright would need to be photographed in good light, with professional equipment, to ensure quality posters and prints. So long as museums and other collectors were the ones to provide the access to the art, the good light, and the physical space for a photographer to work, they expected to be paid and to approve the quality of the reproductions in question, to protect the reputation of the art they owned.

Certainly, this young man could be charged with violating the museum's rule prohibiting photographs, but that could lead to little more than his expulsion. Could he be charged with theft for the act of image-stealing? Or only copyright violation, if he followed through and reproduced copies of the painting for sale?

Legally, these questions are largely settled: charges of theft, a criminal offense, would not plausibly apply. Trespassing might, if he does not leave or returns when asked not to because of his rule-breaking. And the violation of copyright infringement (generally not a crime, but a civil matter) leaves the young man vulnerable to civil suit for financial damages. But while artists and owners of art – along with and their business representatives – can be heard claiming theft when their work is reproduced without permission, it is not the work itself that is stolen, but rather two important qualities of that work that are foreclosed. The fist is the right for the work to be used and seen only as the artist or the owner prefers. The second is the right of the owner to benefit from the money the image can generate.

If I own a work of art, I might use my physical possession of that art to share it with some people but not with others. Especially if I am the creator of that work, I might want the work to be seen only under some conditions and not others, just as I might want it to be seen not in pieces but as a whole, not altered but as I created it. Many countries have specific legal protections of this right of the work's creator – the French call it "droit morale," or the moral right of the creator to prevent a book from being published with changes not approved by its author, or a film not to be colorized or re-edited by the company distributing it.

To the second point, if I can't sell my painting or my book or my film because others are out there selling copies of it, or giving them away without my permission, I'm being deprived of income I would otherwise be able to secure from my talent and effort. That comes closer to theft – I feel my pockets being picked – but even in this case the money lost is hypothetical; it is not something taken but an opportunity foreclosed.

When I make a copy of a song in digital form without payment and without permission, the song too seems perfectly unharmed, still sitting on my sister's hard drive, or still right there, pressed into the vinyl of my mother's old record collection. I didn't run out of her house with her *Barry Manilow's Greatest Hits* under my arm – I just recorded it as it played, and then posted the recording to my personal web page so my friends could enjoy "Feelings" as well. What is "stolen" is not the painting itself, and not the song, but the money that I might have paid to have my own copy of each. Still, the gray area here is quite large. If I had to pay to take a picture of a painting, I probably wouldn't, so there's really nothing lost. If I had to pay for "Feelings," I'd just skip it and turn on the radio instead.

These losses are very different from what we usually think of when we think of theft. If you steal my coat, I'll be cold when I run into the street to try to chase you down. You've done me direct harm by creating an absence. Something I once had, I no longer have. Not so when you "steal" a copy of a picture or a song.

The Digital Easement

Clearly, we need to create some middle-ground on theft in our minds and our laws – and in our machines and databases – to allow, for example, a driverless car to access a proprietary data-base of doctors and hospitals when a passenger in the back seat unexpectedly falls dangerously ill and needs medical help. If the data can be found but requires a payment for access, shall we tell our automobile's operating system to get that data at any cost? Perhaps at some cost but not too much cost? Or by any means legal or illegal, in order to save the human life in the back seat? Broadening the question, is it theft to take the information without paying a licensing fee in order to save a child's life?

The notion from Anglo-American law of the "public easement" describes the use of private property for public good even if the private property owner would rather not allow it. When a power line must pass over your property to reach the local hospital, whether you like it or not the air space you might claim as part of your property, as well as access to it for installation and repair, will be taken for that use. Generally, an easement is more easily defended if the harm to the property owner is slight or nonexistent. Applied to the digital world, we might call the forced access to that medical database a kind of digital easement – something akin to the powerline heading into the hospital.

Not long ago, India did something similar in issuing a "compulsory license" to a very expensive cancer drug developed by Bayer, the German drug company, allowing a group of Indian drug makers to manufacture millions of pills that would indeed save lives, but sending to Bayer only a tiny fraction of the income it lost by having its patents violated, by official government fiat. There was no need to get the details of how to manufacture this privately-owned drug – that data circulates more or less freely online. Only law sets the barrier, and India felt that the law's duty to dying citizens is greater than its duty to the system of intellectual-property ownership it had, until then, supported.

The compulsory license seems to work in this case in part because it is rare, the common good it serves is great and visible, and the harm to Bayer seems slight relative to its larger business. Thinking through similar occasion for compulsory licensing and digital easements leads us to imagine a model in which such licenses and easements would need to remain rare, deeply beneficial to the public, and minimally harmful to the licensor. As well, the possibility of such licenses and easements needs to be widely known, so that owners of property – physical and intellectual – will know the boundaries, or at least that boundaries exist, even if not clearly defined, and will self-police for the sake of the common good and their practical interests.

Finding the reasonable path

Consider again John Locke, and the idea that each individual has a property in and of himself. You can't do certain things with me or to me unless I, as the owner of myself, consent. So, if you want me to lift some bales of cotton, you must come to me with a price I accept to engage my labor because I own myself and you don't. Equally, you can't make me make music, or make me write you a story unless I want to. For a fair price, I might just want to.

But what happens once I've made that music, or written that song? There's an afterlife of my creation which is fundamentally different from the right-now life of myself. The easy answer to the question of that afterlife is that I can make a deal ahead of time about how much, if anything, I should be paid for the fruits of my creative tree. And because I'm no fool, I know that there's risk in whatever options I choose for protecting my intellectual property, a useful phrase for my work's value once the act of creating it is over.

The fan who records my music on his phone is indeed impacting the price I can demand for my performance, and for post-performance resale of my music. The recorder-and-seller is taking something valuable from me, and without my consent, but I should reasonably expect this to happen in today's technology-rich environment. It's close to theft, but no one can say exactly what has been stolen beyond an opportunity to earn an amount we can't determine in advance, so it's a special kind of stealing. It's both a plus and a minus – one more fan in the room, paying for his ticket and making the room feel bigger and my act more exciting, adds value to my reputation and helps a little bit in filling seats at my shows and energizing an audience to seek out my music. This kind of theft cannot be put in the same category as taking my coat, or the physical master-recording that I have physical possession of. It is theft of a degree of the value that I might yet realize, of money I might yet make, and control of my music I might yet exercise, dependent on a hundred or a thousand other variables that I can't control. This kind of violation stands apart from what we mean when we yell "Thief!"

The speed and ease by which information and entertainment travel today are vastly faster than ever before in history, and the costs are vastly lower. That makes individual bits of information and entertainment less scarce than they've been in the past, and thus the market price for them tends to be lower.

The unauthorized copier and distributor of your art or mine is part of this process, and part of enriching the collective experience of what Locke described as that part of life "common to all men" (and we hasten to add, all women too). The shift in scarcity from the copies of the things we produce to the witnessing and sharing of the experience of production itself – from the (trending down) price the musician can demand for the record album or the author for the book, to the (trending up) price for the live performance or writing-for-higher – is sensible. It changes the distribution of rewards and brings more to some creators and less to others, but is hard to see as running afoul of the fundamental values most of us hold, and seems a necessary part of a growing enrichment of the human experience.

Many of us have been discussing these questions for decades now, but they are getting more complex, seemingly by the day. Consider a set of data that describes your commuting habits. This data can provide insights into your life, what you might want to buy and sell, how you relate to your family and your workplace, and whether you drive safely and skillfully.

Combined with similar data from others, it offers insights in whole slices of the population. When shared, this data might be of great value to companies, governments, and other forces in the world about which most of us know very little. Will this data be used in ways that might enrich our lives? In ways that might harm us? In many cases we simply don't know.

Locke's language is helpful to think through the ethics at play here. If my "person" is present in this data, it is mine. It is my property, and the right and wrong of it should begin there.

Chapter Six

Privacy   Did you know your colleague's father is very ill?

Here's a real-life story about a well-intended, sophisticated employer getting things badly wrong in a way that hurts a prized employee and violates strongly held company values. The story illustrates a few of the privacy concerns that most of us pay too little attention to, and which we need to think about a lot more as we apply our values to the emerging digital world. Some details have been changed to hide the company's identity, but the story is real.

The Privacy Advisory Company (we can call it PAC) is in the business of selling (to big companies) good ways to protect customers' and employees' privacy. You'd hire PAC to have a look at how good a job your company does in protecting the information you gather about your customers – including financial information but also things like whom they buy gifts for, what kinds of things they buy – as well as how well you protect personal information you inevitably learn about the people who work for you, ranging from simple things like where they live to more complex information like their family and health status.

PAC is known as a thoughtful, compassionate provider of these business advisory services, and is always rated as a "great place to work." Its salaries aren't the highest but the benefits are great – including generous healthcare programs for employees and their families and extended leave policies for new parents, financial support for care-giving for family members, travel and education. It's that kind of place.

So the irony was thick when an employee named Pat opened her email one morning to see a well-intended note from her boss about her father's health. How sad, he wrote, to learn that her father was entering the final stages of his illness. And what a fantastic person Pat was proving to be – showing concern and spending time with her father in his last days. Or was it his last weeks or months? Either way, her boss told her, she should have no worries about taking the time she needed to help him and the rest of her family.

Pam was stunned. She had not told her boss about her father's illness. She chose not to share that, in part because she wanted to experience her father's passing as a family experience, and she wanted some lines between her family and her work colleagues – colleagues she cherished, but who were simply not part of her large and close family. She needed those relationships to be, and feel, different. And she knew that her boss' kindness could too easily take the form of trying to protect her from the stresses of travel while making time for her father, even if that protection was precisely the opposite of what she wants.

How could he have known, she wondered. And then it dawned on her – the new SmartPal application on their computer system must have flagged her online shopping for books about grieving the loss of a parent, and about father-daughter relationships in their final days. SmartPal was a new tool for mapping the interests of company staffers. It could match colleagues inside the same company who had the same personal passions, or the same kinds of work tasks, but didn't know it. Pat often worked with energy and mining companies, and among PAC's thousands of staffers around the world, someone else might be working on a project for an energy or mining company that would give her a big boost on her own work – if only she knew what her thousands of colleagues were working on and thinking about. SmartPal would silently take note of all these things and build "interest maps" that people could use to find helpful colleagues and receive alerts about work already underway that could be great to know about.

Everyone at the director level and above would also get copied on the interest maps and alerts that their staff members received. Thus Pat's boss came to learn that not only had she been shopping for books about how to deal with a dying father, but also Googling "death of a parent," "Father dying," and "becoming an orphan at 40." In this case, Pat the privacy consultant's privacy was violated – and it was the company making hundreds of millions of dollars helping employers honor the privacy of others that blew it.

The human ethics layer

If we take preservation of privacy as an important principle, we must ask with each new tool and process that touches our organizations and our own actions in the world, does this strengthen or weaken people's privacy? If it weakens their privacy, is that trade-off reasonable? One test of whether it is reasonable is what old-fashioned moralists once called the "front-page test." If we knew the front page of the world's newspapers would share news of our plan to use a new information tool, or our new way of doing our business, would we still do it?

This human habit of referring back to our core principles as a central part of planning our work and lives is part of the digital-ethics mindset. It leans toward openness about new things we do – testing our impact on the world by presuming that the world will know what we do, and that we are the ones doing it. But we have to return to the question of exactly whom we mean by "we." Digital ethics is both an individual exercise and a group activity. At the group level, things often get harder.

What happens when we don't agree on what is right and what is wrong? Certainly new tools that share information we don't have wide agreement on sharing, or execute work we don't have wide agreement on needing to get done, should either not be put into service, or should be built with toggles that allow opting and an opting out of participation for those who might not approve and deserve to keep their autonomy.

The speed of new agreement

But agreement on values is always changing; people's opinions and beliefs change, especially as they learn and experience more, and the people they encounter bring new ideas and experiences to the social conversation about what is right and what is wrong, what is the best life and what is trivia and distraction.

As our tool-development speeds up, we need to speed up the slow-motion negotiation of core values among different communities that has taken place in increments of centuries. When the coming together of different traditions unfolded at the speed of footsteps and animal caravans, and then at the speed of migration by sea and road and eventually by air, we were forced to negotiate change at these speeds.

Today different cultures and communities bump into each other at the speed of electronic connectivity. The results include a flowering of languages, arts, technologies and philosophies. The results also include resistance, struggle, and at times violence. Genuine clashes of belief – for example, the believe that all human live is equally precious versus the believe that one group of people has a higher and more important purpose than others – are the hardest to reconcile. As long as clashes like these remain unresolved, communities are far less likely to share in the value of new technologies and less likely to help others share in that value.

Other kinds of clashes – not of basic principles but of different ways to apply mostly agreed principles, to work through the logic of seeming contradictions as the world continues to change – still involve struggle and at times strife, but are more fertile ground for compromise sooner rather than later.

Yelling "Stop!"

The American writer William F. Buckley described his newly-launched conservative magazine of political opinion in 1955 as one that "stands athwart history, yelling Stop, at a time when no one is inclined to do so, or to have much patience with those who so urge it." That time feels like our time as well, with the mass of rising humanity embracing material gain and work more oriented toward healing, learning and connecting, while furious political movements shout in a different key than Buckley's genteel and well-educated conservatism that the progress of our time exploits the weak as it favors the strong.

Buckley's vision was hardly as flat as his polemic implied, as he himself explained in his many writings over the next decade. He did not want history to stop as much as he wanted the progress all around him to be guided by principle rather than short-term interest. In that, he was a proto-digital-ethicist of a certain stripe. He preferred the clash of fundamental values – the longer, harder fights – but his imperative that we open our eyes to the changes of our world as they happen, and take responsibility not for stopping them but connecting them to what we believe, remains a worthy call and one especially well suited for our moment in history, history which of course will not stop no matter what we say. It will, instead, reflect our beliefs if we can clarify, express, and code them in.

First principles

How to speed up the negotiation of fundamental values? The first and most powerful step is to make those values clear. When Khazir and Ghazala Khan stood on the podium at the 2016 Democratic National Convention to offer comment on the presidential election in the U.S. that year and roundly condemned the Republican candidate – and later president – Donald Trump, Mr. Khan was specific in pointing to the applied values that reflect just this kind of difficult negotiation.

"Have you even read the Constitution, Mr. Trump," he asked with passion, reflecting his family's journey from the fringes of Pakistani society to the height of notoriety in America, through the story of his son's death fighting in the U.S. military in 2004, not far from Pakistan, in service to American interests.

That was a question exactly on point for the consideration of digital ethics: given the extreme disagreement and cultural conflict reflected in that year's presidential election, returning for serious reflection to the U.S. Constitution and the beliefs it reflected would be the essential act of negotiating difference to create and to affirm shared principles. The Constitution itself was the result of years of conflict and compromise focused often on the most heated point of irreconcilable difference, the proprietary of human slavery. The document does not resolve the conflict – a fact that should cause shame many argue, and a fact that certainly caused suffering – but enumerated a process for continual reassessment of the conflict, and reaffirmed principles, including property rights for each individual (reflecting Locke's idea that every man has a property in himself), freedom from unreasonable search and seizure, freedom of speech and freedom of assembly that inevitably weighed against slavery.

Progress in the right direction – but too slow. We can take the work of digital ethics to be a process of speeding up the resolution of those kinds of conflicts, particularly as we design, code and implement digital systems; and a process of reaffirming more generally the higher-level principles that help create more human flourishing and freedoms, but only if – as Mr. Khan affirmed – we affirm those principles again and again, seek to understand them again and again, and make them the center of our political, economic and cultural contests again and again.

The digital revolution begins, outside Philadelphia Mississippi

In the rural south of the United States, many of the descendants of African-American slaves lived strikingly modest lives through much of the twentieth century. Indeed, for many those lives could be confused for the lives of their great-grandparents, slaves on the land living under the thumb of owners and their plantation overseers. A century later some still lived in shacks, under the thumb of share-cropping landowners and the sheriffs enforcing laws and customs meant to keep African American people in their subservience.

World War Two was a turning point for the region. Enough men went off to war, fighting for freedom in Europe and returning to their homes to find themselves without the rights of

free men – their families without the rights of free people – that
some, like E.D. Nixon in Montgomery, Alabama, began
agitating for change (Nixon's work as the leader of the NAACP
in Montgomery led directly to the Montgomery Bus Boycott and
the rise of the young minister Martin Luther King).

By 1964, the prospect of change was palpable enough that
organizations with names like the Congress of Racial Equality
and the Student Nonviolent Coordinating Committee were
quietly working in some of the most dangerous areas for civil
rights advocates – the back country around towns like
Philadelphia, Mississippi. And the efforts were far from one-
sided: the Ku Klux Klan was active, and some sheriffs and
sheriff's deputies (often members of the Klan themselves) took
it upon themselves to mete out punishment to African American
people pushing against the system.

One measure of the climate of fear and domination near
Philadelphia, Mississippi was the finding of nine African
American bodies in a swamp, when federal agents went looking
for three missing civil rights workers that year – two white men
from New York, and one young African American man from
the area who worked with them. Because two of the missing
were white, because they were from families that were media-
savvy and able to rouse the interest of northern politicians, their
disappearance triggered a search. The three had been killed and
buried in an earthen dam under construction in the area, but the
searchers did not know that and so they dredged a local swamp
where locals knew the sheriff and the Klan to dump bodies. And
in that swamp, nine bodies were found, all African American,

none recorded by local law-enforcement as missing, none officially the victim of crime, one body clad in a Congress of Racial Equality tee-shirt.

And it turns out that the rise of a new wave of communications technology was a direct link in the story of how those three missing civil-rights workers – the three that the larger world noticed – wound up near Philadelphia Mississippi and wound up dead. One image captures the new-technology side of the Civil Rights Movement's growing effectiveness during the Montgomery Bus Boycott in 1956. A bit of television news made a few broadcasts on northern television stations and networks during the uncertain days of the boycott's year-long stretch. A young minister – 26-year-old Dr. Martin Luther King, notably youthful and slight of frame – stands, wearing a suit and hat, being interviewed by a reporter holding a microphone, man-on-the-street style, sharing his view of the boycott's progress. Most notable is that in 1955 and 1956, there were only two commercial television stations broadcasting in Alabama, both with tiny audiences, on-air only for a limited portion of the day, and almost impossible to pick up in much of the hilly state, which barely mattered, with so few televisions in Alabama.

The import of the televised broadcast of the young reverend's perspective on the world-historical bus boycott had almost nothing to do with the few who would see it in Alabama, and everything to do with the elites – those few with television sets of their own – in New York and Washington DC and Atlanta and Chicago and Los Angeles, able for the first time in history to see and hear the people and places in our own country living under laws mandating thoroughgoing racial domination by white people over black people. And ten years later, when local clergy and activists led a march over the Edmund Pettus Bridge in Selma, Alabama, and CBS flew footage captured on the scene of state police converging on the peaceful marches and beating them badly (including future Congressman John Lewis), killing one woman in the process, the network made history by bringing police warfare against civilians marching peacefully for redress directly into the living rooms of millions of Americans.

The repression and violence were centuries old but the technology was new, for the first time only in that new historical moment able to carry it into people's lives in other cities and towns across the country. And with that vivid awareness of what the violence and the struggle against it looked and sounded like, the nation's conscience was moved as it had not been before, on this issue born with the very foundation of European settlement in the Americas.

When smaller places are connected to larger places, made more visible and woven into the common experience of large numbers of people outside and far away, the values of the larger society tend to migrate toward the smaller towns, local cultures, individual homes and even into the pockets of people young and old, where our smart devices now nestle.

There is a loss here, but also a gain — if the values of the larger society are ultimately values that support the dignity of all people and shared human abundance. It's that "if" which matters so much and must be the cornerstone of digital ethics.

Chapter Seven

Fairness

Recall the example of a "smart gun" discussed earlier. A police officer points his weapon at a person a few feet away and a visual sensor at the tip of the gun captures an image, runs it against a facial-recognition database and sends a signal to the officer – a certain pattern of vibration, or a sound – to note that the person it's pointed at is, say, a convicted violent felon, or an under-cover police officer, or deaf, each of these possibilities changing the way a well-trained protector of the peace will likely act.

There is not only intelligence built into that smart gun, but history as well. The collective data taken together tell millions of stories about millions of people, and stories about who winds up in in jail, who winds up on a police force, who is likely to have clean criminal record even with a rough-and-tumble personal history, and who is likely to have been given no breaks by the law. These stories go beyond each individual and include the history of their families, their neighborhoods and their neighbors.

The child growing up among violence, without parents in the home, speaking a certain language, looking a certain way, cheek-by-jowl with frequent law breakers or nestled among doctors and lawyers and judges – will this child's story be told by databases in ways that reinforce the unfairness of circumstance and social history? Factors ranging from pure economics to red-lining lenders and keep-them-out neighbors to the good-fortune of being raised by a protected and protective family will shape the data that will tell decision-makers who can be trusted, who can be given a break, who is dangerous and who is safe.

Mr. Loomis Goes to Jail

Some courts in the US are already using AI-driven sentencing tools to predict how likely a convicted offender is to commit future crimes if released or given a light sentence. The tools are driven by a growing range of data, often including performance in school, the level of crime and criminals near the to-be-sentenced offender's home address, his or her criminal history, work history, and more. At least one important case – *Wisconsin vs. Loomis* – raised the question of due process in based on these kinds of databases and decision-tools.. In 2013, Mr. Loomis was charged and eventually convicted of committing a drive-by shooting in the town of LaCrosse. He was sentenced to six years in jail. He appealed his conviction to the Wisconsin State Supreme Court, and eventually to the U.S. Supreme Court, on two grounds. The first was that some of the inner workings of the AI sentencing-recommendation engine hid the factors that determined the sentence, and the second was that both race and gender are in fact among the visible elements that the system, called COMPAS, does take into account in its algorithmic work.

The Wisconsin Supreme Court upheld Loomis' conviction, and the U.S. Supreme Court rejected the request by Loomis to have his case heard there, effectively affirming the Wisconsin court's ruling. Mr. Loomis remains behind bars.

John Roberts, Chief Justice of the U.S. Supreme Court, is certainly aware of the challenge cases like *Loomis* represent. A 2017 article in the *New Yorker* magazine described an exchange he had with the president of Rensselear Polytechnic Institute:

"Can you foresee a day," asked Shirley Ann Jackson, president of the college in upstate New York, "when smart machines, driven with artificial intelligences, will assist with courtroom fact-finding or, more controversially even, judicial decision-making?"

The chief justice's answer was more surprising than the question. "It's a day that's here," he said, "and it's putting a significant strain on how the judiciary goes about doing things."

Sue's daughter: special treatment?

As a third example of new kinds of fairness questions that we are only waking up today comes in a classroom. It is less overtly extreme than cases issuing from the business end of firearm, but in the end perhaps even more consequential.

Eight-year-old Sue is a remarkably smart individual. Her parents know it but try not to brag, and they don't want her treated too differently from other kids. "I want Sue to know that she's a fabulous, powerful young lady," her mother says, "but not just because she can score the highest on a test or memorize a poem. I want her to connect with other kids, to see the value in community, and understand that her talent is something that gets stronger when she connects with other people, and it's something that can help other people reach their goals too." Still, at the age of eight, Sue is clearly gifted: she can solve complicated math problems (her parents haven't gone beyond multiplication, but suspect she can do a lot more), and she enjoys memorizing story books and poems.

"Still," her mother says, "I want her to be with regular kids who have talents she doesn't have, kids who might be average on some tests but lovely and charming and interesting in their own right." So Sue does not go to a special school or join in special after-school programs for "gifted" children. In the third grade, she does mostly the same school work as her peers, though she's in the advanced math class for her grade, and often memorizes short stories and poems she and her classmates are assigned to read.

Then one day Sue comes home clearly excited. "Mom, mom!" she says, "I'm a diamond!" Her mother's confused, but after smiling along and sharing her daughter's excitement, she learns that as part of Sue's school day, every student in her class took a "dynamic, graduated assessment" online. That meant that each student spent about 20 minutes in front of a computer screen, answering questions across a range of academic topics, using a software program that adjusts to a student's level of ability – so most students would be seeing different questions from the rest of the group, questions specifically geared to the level of their ability. And Sue pretty much broke the test. She answered all the math and logical reasoning questions right, well into the high-school range at the top of the program's assessment scale, and demonstrated college-level language skills. "My teacher told me that I have the highest score ever! Better than gold. A diamond!"

This was exactly what Sue's mother had been afraid would eventually happen. Sue was now identified as a super-performer and faced some degree of separation from her friends. "I don't want her in a special room with special kids, having a sense of being different and superior reinforced." Sue's teacher was surprised by that reaction. She stressed that with new computer-based learning tools, every student could be reached at exactly the right level of challenge, in a way that one teacher standing in front of a group of twenty or thirty students just couldn't pull off. But that kind of compromise, of being part of the group even if it meant working on easy material better suited to the average students in room, was exactly what Sue's mother thought her daughter would benefit from most, at least for a while longer. In high school and college, she thought, let Sue be the genius she is, let her work with professors and win prizes and advance knowledge. But for the next few years, let her just be a normal kid.

"I've read about these sports and music academies in some countries," Sue's mother says, "where all the kids in the country are tested when they're nine or ten to see who has the most amazing natural ability, who can jump the farthest, who has the longest stride, who has perfect pitch, and then they're taken to these special sports and music schools, boarding schools, and trained to be the best in the world to compete internationally and win glory for the country and for the king. And you know, that's the last thing in the world I would want for my daughter and our family. I think it would be terribly lonely for her, and for us. And I think she'd miss the chance to grow the other dimensions of who she is, to do things she's not so great at, to come in second or fifth or fiftieth, and to learn to be part of a community where sometimes you're the best and sometimes you're pulling up the rear, and the other virtues you need to exercise like kindness and caring are part of what you learn too."

Sue's mother had been sheltering her daughter from this kind of assessment and "differentiation," but now that every classroom – every computer, even every phone – can deliver assessment and ranking, she'd have to try even harder, or maybe rethink the whole project. "Maybe in today's world," Sue's mother says, "it just can't be done, and we have to get used to being scored and sorted. Maybe it's OK to know that Sue ranks as a diamond, and maybe I'm just two stars as a mom. Maybe

Sue will benefit from clustering more with the other diamonds out there."

Losing the cosmopolitan virtues

We all live in this new world already, at least to a degree. Most of our children are sorted by their school tests and assessments to at least some degree, and when that sorting is done well, it often means that students do in fact find themselves working on challenges most appropriate to their skills. But when the sorting is less than ideal, it can become self-reinforcing, as a bright student relegated to a dim group gets used to underperforming. And even when the assessments are spot-on, we do indeed lose some of the cosmopolitan virtue of the mixed group, of the simple-minded solution that happens to solve the brilliantly complex problem.

That's only half of the sacrifice of predicting winners and making our investments in education – or our risks in underwriting insurance, say, or deciding who is allowed to drive a car or fly a plane – based on automated assessments, no matter how good those assessments might be. When we limit the number of happy accidents and paradigm-busting end-runs around the challenges of mind and heart, we reinforce our own theories and that means more predictability, more accomplishment in a straight line, more and more-efficient attention paid to the known challenges in our lives, and less and less discovery of unexpected problems, strengths, challenges and joy. It also means less "leaping," as the poet and critic Robert Bly describes what poets do when their work seems off the proper path of logic but nevertheless deeply true, perhaps like

Handwritten margin note (right): * I can't wait till my doctor/Pilot etc was trained!

Handwritten note (bottom): Schools have always differentiated between years and separated by class and class teachers differentiate between pupils. Sorting is NOT new.

Emily Dickinson's observation that "Hope is a thing with feathers," or e.e. cummings declaring that "no one, not even the rain, has such small hands."

We lose the virtue of messiness with more skillful evaluation and prediction – fewer rule-breaking breakthroughs, and fewer moments of realizing that we, the rule makers, might just have it all wrong. Chaos is a high price to pay for these kinds of virtuous accidents, but bearing in mind the need for a hedge, for a slice of something more chaotic or open-ended in the pie of prediction, is a worthwhile practice.

Technology-innovation pioneer Alph Bingham calls part of what's going on here the power of "optimal distance," and it's similar to the underlying mechanisms of evolution. When we match individuals to tasks based on our assessment of who is best at what, and what level of challenge serves given individuals best, we create a system in which our champions and elites are drawn from the predicted best. We lose the virtue of the surprises that come from unexpected strength and skill emerging from those seeming to be without them. This kind of surprise – which we can find throughout our lives as well as throughout History writ large – is something we should not, and one can argue *cannot*, do without. The loss would be felt by the individuals who might just perform remarkably well though our predictive apparatus misses this likelihood; they would have no change to defy expectation. The loss would also fall to the group that accustoms itself to a lack of surprises and lack of hope for personal accomplishment beyond the lines of the predictive box. We would all risk trusting too much in the predictions and evaluations that might come to narrow the range of our experience and contributions – and leave each of us less surprising even to ourselves. Our culture generally and our enterprises in particular benefit enormously when unexpected approaches to known problems, hidden paths to goals, and reframing of challenges from unexpected perspectives emerge. Particularly when the seemingly unskilled or unable win the day, our notions of how the world might work and why are shaken – and we are prompted to become larger, wiser and more humane.

Anthony Appiah's tightrope

Philosopher Anthony Appiah has been thinking through what "cosmopolitanism" means for the last couple of decades, and has offered helpful insights about how important this broad idea is to the emerging digital age. What Appiah calls "that new interconnectedness," is, he explains in an interview with the *Mandela Journal* at the University of Georgia, "the thing that makes cosmopolitanism absolutely the right philosophy for the present because we need a way of thinking about people that we are connected with through communication that allows us to respect them and to be concerned for their welfare but doesn't lead us to try to force them, in illegitimate ways, to be like us. And again it's the combination of universality plus difference that marks off cosmopolitanism from some of the other universal views." Capturing that universality in the form of an ethics layer for each of our digital systems is the challenge, particularly in the task of having it do what it must but not more. That's allows Appiah's "difference," as well as universality. Cosmopolitanism, he explains further, is the coming together of two ideas: "some form of commitment to the universality of concern for all human beings. . . .that sense that everybody matters, every human being is important," as well as "the idea that people are entitled to live lives according to different ideals, different conceptions of what they're up to, what they think is worthwhile." This is the tension we feel, the tightrope we walk across as we practice digital ethics.

Avoiding a smaller pool of problem solvers

Appiah celebrates the social and civic experience of different communities coming together and living lives made richer because of diversity and difference. That diversity and difference support contribute to a society's ability to solve problems better and faster – a very good thing.

As a general rule, proven methods of solving problems are effective with familiar problems. In a world filled with change though, a wider range of approaches means better resilience for societies and more chance to thrive. In leaning on familiar strategies and familiar problem-solvers, we lose a lot – including the angles of vision that sometimes help us see more. We lose the blunt-force genius of the novice whose amateurish idea turns out to be perfect for a distinctive and novel challenge, or whose mis-perception of a problem neatly overlooks exactly its most misleading element.

Breaking habits-of-mind to solve problems.

Try this exercise. Pull out your phone and open up any of the games that come programmed into it. Most will involve pressing buttons or tilting the phone or in some other way moving an image on the screen around to stop an attacker or build a wall or complete a circle. Take a minute to play the game – these games tend to have short cycles, so you can probably play it a few times pretty quickly.

If you think about how the game works, and apply your intelligence and skill, you can probably get better at it fast. But if you try, instead, not to think – just to play it, and lose, and play it again, and don't think, and play it and play it and play it, you'll most likely get a little better at it with each attempt, without any conscious reflection on the rules or on what you're learning. Something's going on outside of your conscious intelligence that's a lot like what the low-scoring student can contribute to a complex problem that she "shouldn't" be able to solve.

Get enough of those low-scorers together and give them enough attempts and you'll likely see some interesting and useful solutions emerge, most taking different approaches than the fully qualified students take, which means the expanding of the frame of the solution set that only the poorly-qualified tend to accomplish. There's a wonderful scene in the television show *The Wire* that captures exactly this. A group of undercover detectives are trying to crack the code for a signal sent from payphones between low-level players in a drug gang. It turns out the street dealers use the location of the numbers on the keypad, not the numbers themselves, or the corresponding letters, because they can't read, and numbers make them nervous. They remember the code because it's a cross, or a box, rather than a sequence of numbers or a word that verify their identities and close their deals – and it's a code that's almost unbreakable because the dealers don't use the written language of the police chasing them. Their difference – their deficits – make them geniuses in this specific circumstance.

Fairness and the feeling of fairness: How large is your set?

Predictive evaluation – judging the criminal's likelihood to commit other crimes in the future or judging the young woman to be Ivy League material or worth the investment of special classes and training – always comes down to the question of the size and quality of the set of data that predictive tools work with. We have to make it a habit to ask: What's the data this prediction is based on? What biases went into creating this data set (a question in high relief looking at criminal sentencing tools)? How long a period of time should we consider, when we consider whether a way of organizing a workplace or a school or even a car full of kids or colleagues on a long trip accords with our beliefs.

All the options for organizing space and resources reflect values; we need to cultivate the skills and habits of seeing those values.

Consider an airplane filled with passengers. Some are better customers than others, judged against a range of criteria. Some are willing and able to pay more, or to fly a lot more often – so they go up front and the rest of us walk past them to the smaller, noisier seats. This invariably creates a feeling of unfairness, in part because the airline is evaluating and sorting passengers based on a large set of data, the large set presumably most relevant to the airline's prosperity.

Fair and unfair *right now*

Based on a month or a year or decade, it's fair from their perspective to give more rewards and a better flying experience to the passengers who will spend lots more money with that airline over those longer periods of time. Indeed, based our own lifestyles of frequent, long trips, the idea that flying 100,000 miles a year on an airline might earn us occasional upgrades seems eminently fair. Yet a man or woman – and especially a child, in our experience – walks past normal looking men and women sitting in big comfy chairs or even reclining in flying couches on long-haul trips, and immediately feel the unfairness. "What makes this person better than me," we often ask as we instantly size up the situation, and we don't mean what makes them better customers. We mean that under the shared duress of a long flight, when we all feel a taste of danger in the high-altitude air, and we all must find a way to share limited resources in a small community flying far above the nations and laws that govern us most days, what makes this one person better than me *right now?*

— Just plain silly

"What kind of place is this?"

The problem is an example of set theory: the divergence in customer treatment might be sensible and even fair if we take as our basis of judgement many weeks and months, but in the moment, we have an almost instinctual sense that the only time that matters is *now*. Treating a few clearly better than the rest just seems patently wrong in that moment. Another way of putting this is to observe that airplanes – and buses and cars – are at once means of getting to a place, but they are also places in themselves. The question of "what kind of place is this, and what kind of place do we want it to be" is the question too often left unasked.

The need to know more

The set-theory aspect of this kind of ethical thinking is especially important in the age of algorithms and predictive analytics. When we ask, "What kind of place is this?" how can we consider the limits of the data we use to frame the answer? For example, if one factor in whether to offer a man accused of a crime the chance to be released from jail while awaiting his trial is where he will live, what the crime rate in that specific place is, and whether people with criminal records tend to cluster nearby, we might rely on a database to reveal these facts and determine – as programs built to guide in these decisions often do – that the risk of this person committing crime while on release is too high and so he stays locked up.

Yet what if we know more about these places – if we add more years to the range of time we explore, and more range to factors that have shaped zip-code-level frequency of crime and density of criminals? Does it matter, for example, that both public and private landlords had kept people of certain races out of bordering neighborhoods, at times through rental policies and at times through laws openly applied? Does it matter that the home of a man on trial – eligible for release, if the algorithm gives the nod – is in one of the few neighborhoods where for 20 years police practiced "stop and frisk" searches on just about every young male with great frequency, leading directly to more arrests based on the same facts and behaviors compared to most other neighborhoods nearby? Expanding the set of data to capture these patterns, influences and social meanings can help increasingly powerful tools resist and reshape old biases, though the work is challenging and might, on the surface, seem unnecessary.

Shaking, sputtering versions

New technologies generally begin in shaking, sputtering versions of what they will eventually become. Early automobiles, early airplanes, and early versions of now ubiquitous software applications failed to turn on, stay on and do what they were built to do as often as not.

The physician keenly in need of five miles of swift transport down the road to help save a dying child would be mad to use his early combustion-engine automobile – would it start? would it sputter to a stop on the road? – when his perfectly well-fed and harnessed horse just stands idle in the barn. It's just too soon for the automobile. The pre-World-War-One pilots knew they risked death with every exhilarating test flight in canvass-winged biplanes, and many of us still walk the earth who brought newspapers and magazines over to their desktop behemoths because signing on to the early-days internet and – even bolder – trying to download an image file would take long minutes of idle time. But we understood that better days were coming. Always-on connectivity, globe-spanning jetliners, safe and reliable automobiles by the tens of millions – all were on their way, starting with these virtual dinosaurs suited mostly to experimenters and thrill-seekers.

Earl-days equity

In the early days of a powerful new technology, equitable access to it is hardly a compelling thought. Do *more* people want to fall out of early airplanes into the sky? Do we owe it to all our neighbors to install access in their homes to DARPANET's high-security, imageless cursor prompts? Not when these technologies do relatively little good for most users, and some real harm to at least a few. But as they mature, as these technologies become reliable, consistent, and meaningfully rewarding to almost all who use them we have recognized the importance of common access to once-new technologies like electricity, clean water and public roads.

In much of the Western world, the development of new drugs offers a powerful view of the ethical challenge of developing new technologies with the potential for broad value, but also with potentially great cost and risk. The model for access to new pharmaceuticals we live with today addresses this challenge in a highly structured way that leads to broad common access and great common good, *eventually*.

In the U.S., new drugs undergo extensive and expensive scientific study after they have demonstrated potential efficacy but before they are available to the public. During this multi-year process, drugs created by for-profit companies generally begin a 20-year period of patent protection, though the years of the trials eat away at that period. The protected period in the market ally from 12 to 18 years. After that, other drug makers can make "copycat" versions and sell them at whatever – generally much lower – price buyers are willing to pay. So, during the protected period, the price of some drugs is quite high, even unaffordably high for many, while eventually most drugs become modestly priced and widely available.

The patent protecting of drugs is perhaps the most stark example of the time-dependency of privileged access to technology. The computers costing millions in the 1960's were generally less powerful than the computers in a decent $30 digital watch today. In 1950, the cost to carry one passenger across the country in a commercial airplane was several times it is today. Early cell phones cost several thousands of dollars in the 1970's, weighted as much as small bricks, and barely worked.

The social imperative: the greatest good for the greatest many

So at what point do we look at technologies like life-saving drugs, the electricity grid, or the global internet and declare that this is no longer a special luxury or a tool for well-funded industry but now a proper part of ordinary life for ordinary people? Time is a central factor in this question. Answering it should begin with a commitment to the greatest good for the greatest many from new technologies and will also require a plausible sense of how and when costs dramatically decline. This might take a form – like ensuring incentives for new-drug development through patents – that could seem cruel and needs constant modification and exception-making. It might take an opposite seeming form, like government mandates for low- or zero-emissions cars, which spurred the capital investment needed to speed the development of new kinds of energy deployment for engines. In all cases, though, thinking through how best to ensure the broadest benefit to the broadest user base is a social and ethical imperative.

* this depends on who is paying for the AI programme!

Rooting the *digital* ethics experience in *human* ethics

A group of women in a small village in a rural, low-income country feels lucky. A woman from a more-developed country has been visiting, guided by a professional from a nearby city (still two days of walking from the village), to help start a lending circle.

When digital technologies help support new ways of living and doing business, should these tools have built into them a sense of social justice? Should they allow the strongest individuals to maximize their personal interests without regard to the interests of others – and perhaps increase inequality even as they increase living standards? Should they be able to empower local leaders in new ways that might be abused, or should there be controls built in to limit – or report on – the actions of the new winners in the new economic prosperity?

Consider the case of microfinance in very poor countries. Very low-cost digital communications technologies have helped foster a wave of lending that touches millions of lives and extends new opportunities for poor people to start enterprises, create value, and earn money for themselves and their families. Good things, no doubt.

The typical microcredit operation works like this: instead of demanding that a borrower have a business track-record or substantial collateral, the model allows "unbanked" people with no assets to borrow enough to start a small business, often as part of a local lending circle. If one individual fails to repay the money lent, the rest of the members of the circle lose their ability to borrow more. This creates a kind of social collateral, and incentives whole communities of people to help support new businesses and ensure that loans are repaid. But it also incentivizes these local lending circle to intimidate and threaten slow-payers. Is that OK? Using technology to spread economic power down the pyramid of wealth also spreads the potential for abuse – unless the technology tools themselves build in controls of behavior and local power as well. But should the tool-builders encode their local values in tools to be used by people far away, from other cultures? If the answer is yes to a degree, what is that degree? What are the inviolable principles that should be coded into the tools that help support lending?

Pioneering microlender Grameen Bank in Bangladesh made women receiving loans promise to limit the number of children they would have. If that's seen as OK because having fewer children predicts higher likelihood of being able to repay loans, what if being a member of one ethnic group also turns out to be predictive of likelihood to repay? What are the limits of pure utility in setting these kinds of policies?

Here's how a microlending circle often works: five women in a village will each be given about $50 – more than two months' worth of income for the average person in the village – and will be asked to use that money to support an informal business that can generate enough income to pay back the loan with interest, and help each new entrepreneur and her family sustain themselves and improve their lives. In the early days of microfinance in central Asia and Africa, these businesses often brought basic goods and services to very underserved local markets. A woman might buy a cow and sell the milk; in many villages, no one had the capital to invest in the cow, though milk would be readily bought if available. Many women became "phone ladies," buying a mobile or satellite phone and paying for monthly service, then selling access by the minute to area residents with family in distant cities or far-flung villages that had their own phone ladies.

By 2010, more than $40 billion was circulating, mostly in low-income nations, through microfinance institutions. These loans actually tend to have high interest rates by developed-nation standards, often between 10 and 30% annually, but that's generally much less than rates charged by traditional "informal" lenders in poor-nation villages. The best point of reference for these lenders would be loan-sharks and other lenders ready to bear high-risk clients with physical threats the unstated motive for repayment.

Most microlenders rely instead on two key dynamics to keep their repayment rates high and their operations viable. The first is a very high threshold for who qualifies for a micro loan. Standard measures of creditworthiness aren't useful in small villages in poor companies – or in the massive shanty-towns and *favelas* surrounding large cities on several continents – where no one has a credit rating, where few if any formal banks operate and virtually none of the residents have ever had a formal banking relationship. In these areas, because geography and large-scale, structural economic gaps have kept almost all people out of the formal lending and banking system, even the most disciplined, hardest-working people have never had formal loans. Gender in large areas of the world plays a similar role. In some countries and in many local areas, women are simply not included in the formal economy. Even women with substantial assets or income are not extended loans by banks.

Microlenders coming into these environments often seek to serve the elite of these excluded populations. As the CEO of a large microlender in Pakistan described it, "instead of looking at a credit history or a formal documentation of income and assets, as maybe we can do in the larger cities here for a few people, in the villages we create a profile of the people we might lend to, talking to neighbors, looking at how they keep their homes, whether they send their children to school, whether their children are well looked-after."

The other major tool to help ensure repayment of loans in local areas without traditional banks and banking habits is the lending circle. The micro-finance institution might not be able to encourage a woman who is late with a loan payment to pay up using the tools an impatient lender in New York or London might use – threats of a poor credit report, formal and legal demands for payment, perhaps even a lawsuit. But if your neighbor cannot get a loan to buy a cow to make the money she needs to feed her children because you have failed to repay your loan, the social pressure you'll face is likely to be intense. If your lending circle includes family, friends and local leaders, the pressure to pay will be inescapable. And it might include intimidation of all kinds. Think back to the loan shark as a reference point.

Even allowing someone into the lending circle in the first place calls up a range of biases that might make sense from some perspectives but which from others would simply seem wrong. If in my village, people from a certain caste or ethnicity or family background are excluded from the main life of the community, of course I won't want them in the lending circle because social restrictions make them less able to repay their loans – and I might well have imbibed a lifetime full of prejudice against them.

Lending circles are in fact declining as a microlending tool. Pioneer Grameen Bank has mostly chosen to leave the model behind, at least in its most explicit form. Grameen still requires participation in "small homogenous groups," and the bank's website reports that "The interest rate on all loans is 16 percent. The repayment rate on loans is currently – 95 per cent – due to group pressure and self-interest, as well as the motivation of borrowers."

Research from the World Bank suggests that membership in a lending circle in many cases does not meaningfully increase or support repayment rates; more significantly, as microfinance organizations have grown in scale, more and more on-the-ground staffers have witnessed the ugly side of lending circles, and this informal knowledge – seldom discussed openly and almost never written about – has had an effect on policies.

These informal practices that law generally limits or excludes in the formal economy create an enormous ethical challenge as movements like microfinance bring large populations from outside our formal economies into the world of borrowing, lending, payment and repayment. Informal communities create informal and often unstated rules. Bullying, ethnic turf-building, the reinforcing of local, abusive hierarchies – are all likely bad outcomes.

An ethics layer is clearly needed in the operating stack here – both on the digital and the human-networking side. The blend of smartphones, financial tracking and electronic storytelling that help microfinance spread among people uplifted by the opportunities it brings and the person-to-person, on-the-ground community building need more than an eye toward outcomes. The process, up and down, needs to build on values about the value of human lives, the importance of visible fairness, and equity – just like the driverless car or the smart-city connected network.

Freedom and Innovation: what digital systems help make happen

"My father," a friend tells me, "was an Irishman in Northern England in the 1950's and 1960s and faced extraordinary limits on what he could do. He could not get the education he wanted to get, he could not have the job he wanted to have, and he could not do the work he wanted to do." This woman, the Chief Innovation Officer at a global insurance company, adds that the rise of Ireland to an economic powerhouse – the "Celtic Tiger" economy of the 1990's and 2000's, before the global economic decline that it had more than its share in precipitating – "could not have come about, could not have brought most members of my family from living without hot running water and indoor toilets in their homes to a truly modern and comfortable way of life, without the unlocking of the talent and the creative force of the Irish Catholics who were given a new kind of freedom by the law and by their society in the 1970's and 1980's. Their labor and their ambition were freed in important ways, and led to a deep enrichment of the common good."

And another story: A Jewish man in Austria runs a pharmacy in the 1920's, a shop that sells goods of all kinds including medicines that might be found in any of a dozen other shops among the neighboring towns. But he's a bit of a chemist, curious and creative, and he begins to compound new medicines, blending and refining the commercially available drugs and traditional naturopathic cures his customers count on. One or two prove remarkably helpful, and he considers how we might increase the scale of his operation, become a pharmacy with greater reach and impact – or even a small pharmaceutical company. Across German-speaking Europe, dozens of other chemists are, unbeknownst this man in his modest Austrian city, forming a wave of pharmaceutical innovation. His timing is excellent, on the one hand. And on the other, it could not be worse. The culture bears down against the Jews. Hitler invades Austria, and most of his fellow countrymen are delighted. His shop is taken from him. Along with his family, he perishes in a Nazi camp.

In the American writer Howard Fast's book *Freedom Road*, a young black man, the children of slaves, leaves the United States for medical school in Scotland during reconstruction. He returns to his Southern home and finds himself in the middle of a violent struggle over the status of black men and women – no longer slaves, but certainly not fully free. A group of white vigilantes takes advantage of his training, and forces him to treat their wounded. Then they kill him, because he represents exactly the kind of elevation of station and skill they are dedicated to keep black people from achieving.

All three stories make plain one kind of connection between freedom and innovation. It is direct and unambiguous: people from every station, of every race and ethnicity, from every neighborhood and nation, are drawn to create value for themselves, for their families, and for their societies. Not all people, and not all in equal measure. But innovation of all kinds has been spurred through human history by access to the common culture, to the public square, to schools and to markets.

As we reflect on what we believe and why, and as we wire these beliefs into our machines and devices, we make the world better when we ensure the freedoms of the individual to life, liberty, health and property, as John Locke wrote hundreds of years ago.

Our machines and systems help us do this so long as we, the people who make and run these systems, keep our minds on the virtues that drive our sense of the good and decent life, and our hands on the controls.

Some of these observations peceeding are very well meaning but make the same error that Kant or Marx could be accused of. That of assuming that people will do the 'right thing'. There is no greater reason to assume that people will abandon self interest when programming AI any more than they do when living their own lives.

I suspect that machine morality will reflect our own; with all its flaws and contradictions.

30 July 19

Part Two

Three Documents

to prompt discussion and guide action

In April of 2018 a group of about forty people got together to begin work on crafting three documents – documents described a bit in the first part of this book, and addressing the specific challenges identified in those pages.

The people gathering were technology executives, professors, consultants, HR leaders, graduate students, diplomats, writers and consultants.

They all participated as individuals – not officially representing their employers – but they brought experience and perspectives into the room from their work at engineers, programmers, lawyers, marketers and policy makers at some the largest technology corporations in the world, as well as hospitals, the US government, small consulting firms and one American Indian tribal nation.

This varied and highly-engaged group spent two days that April in 2018 working in small teams to craft initial drafts of documents meant to capture the principles and aspirations that should guide digital development of all kinds; the different roles and responsibilities that should be recognized based on the role we play in the digital ecosystem (the role of creators, of corporations, of users, of governments, and so on); and a brief practical guide for architects and coders of digital systems to make choices as they work that can best lead to a clear and consistent ethics layer is present and effective in every system. We got together again for two days in February of 2019, with a good bit of work in between these two sessions.

We hope that these documents can each be the basis of ongoing conversation among people from all walks of life, with special relevance to technology creators, corporations and other large organizations that deliver and support digital systems of all kinds, and governments and other groups that shape and support standards relevant to digital systems.

The first document – the statement of aspirations – is not meant to be a set of rules, but instead a set of questions and hopes that all of us can and should reflect upon as we plan, use, and improve digital systems. It offers some context to help explain why these aspirations are so important, and why they are legitimate in a world in which we can easily ignore the collective actions in the past that have created platforms and opportunities for our work today. That's why it begins by noting that "digital systems are the result of many generations of work and investment. They are keynote achievements of our age."

The investment of enormous amounts of public money built the early internet and other foundational digital technologies, beginning with the military-science arms of the U.S. and British governments during the Second World War, extending through the technology breakthroughs at NASA in the run-up to the first Earth-orbiting satellites and moon landings, and continuing through the many billions of investment through the Defense Advanced Research Projects Agency, which crafted the early versions of the internet. Time and treasure were taken from hundreds of millions of ordinary people over many decades to fund the platforms upon which just about all new digital technology creation stands today. That's why the ordinary man and woman, and the collective voice of consumers as well as creators, need to help set boundaries and share aspirations for digital technologies. That's why shared human abundance and richer shared experiences for the many, rather than for the few, are the highest goals of technology.

STATEMENT OF ASPIRATIONS

Digital systems are the result of many generations of work and investment. They are keynote achievements of our age.

Many of these systems rightly accomplish specific tasks and reward individuals and distinct groups of people, yet their highest purpose taken as a whole is to create shared human abundance, guided by the commitment that all human life is equally precious.

With this in mind, these fundamental aspirations should guide decisions and policies related to digital systems:

1. The right and the duty for humans to set limits to digital systems are essential.

2. Access to technology, information and learning should be as broad and democratic as possible.

3. Digital systems should maximize human learning and aid understanding of the world. Knowledge should not be hidden but made visible to the fullest extent possible.

4. Personally-identifiable information (PII) should remain visible to and controlled by source individuals and used by others at the discretion of the source individuals.

5. The right to anonymity is essential, with reasonable accountability for impact on others.

6. A high regard for future human generations is paramount. Children need to be a part of the conversation about digital ethics, and our systems and actions need to consider impact on children and future generations.

*

The statement of aspirations is something that digital makers and users should refer to again and again to help remind themselves of the purpose of digital systems and the effects of digital systems that we most want to ensure. This way, we can continually make the many small adjustments to keep on track, and align with our goals and beliefs, just as a sailor keeps an eye on the dock as she heads into port, making big moves and small to keep on course even as winds shift, other boats swing near, and the natural habit to drift this way or that works its influence. The statement of aspirations is the reference point, the true north that we check and recheck to be sure we're heading where we want to end up.

The second document, outlining duties of different parties playing different roles in the digital-ethics ecosystem, has a different ambition. Instead of outlining ideas and aspirations, it talks about who should do what when creating, selling, managing, or using digital systems.

It's more likely to change over time, and requires a greater degree of precision. Like the U.S. Constitution in comparison to the Declaration of Independence, it is not poetry but prose: not the vision, but the rules of who does what. The version here is a starting point for discussion, argument and experimentation. It lacks, in this version 1.0, any kind of mechanism for enforcement. The Constitution establishes the role of the courts in the U.S., and the Supreme Court in particular, to decide whether the actions of the other branches of the government are consistent with the Constitution itself. It's a closed system in that sense – the document establishes an enforcer of the document's integrity. This digital-ethics document does not, but if and when a meaningful number of technology creators, sellers, operators, users and regulators sign on to follow it, adding some tools for oversight will be an important next task.

DUTIES OF DIFFERENT PARTIES

For individuals. Digital citizenship includes a duty for individuals to stay reasonably informed and engaged.

For individual creators. Be aware of the work of others that your work stands upon.

This might be the decades of work of many thousands to build platforms like the publicly-accessible Internet; it might be the code-base that programing languages and applications are built upon; it might be a publicly-built and maintained system of highways and roads that an e-commerce enterprise relies upon.

Reinvestment in, reinforce, and strengthen these commons-like resources.

Be aware of the use of your work by others, and how your publishing, selling or otherwise making available that work will affect others.

The work and it uses would not exist without your making it available: you must take reasonable steps to encourage positive use and reduce harm from what you create.

For corporations and other businesses. Corporations and other businesses must remain informed of and engaged with the sources and the effects of the digital systems they use, create and circulate.

They must compensate creators and other prior owners fairly.

They must seek and secure permission to use personal data and aggregated data generated by people, from those people.

They have a duty to share the realized gains from this data with the data sources in reasonable measure, as well as sharing active participation in as-yet realized gains.

They have a duty to consider and protect against the unintended consequences of their products.

For governments. In a digital age, the fundamental role of government has not changed.

Governments exist to protect the people and promote the common good of those they govern, as well as to show regard for the well-being and human rights of those beyond their borders.

Governments should protect rights and privacy through sharply focused and appropriately limited privacy regulations.

Digital practices in the execution of government work should protect privacy and enable freedom of expression.

Governments should actively support the development of utility-scale platforms that empower individual, local and private creators and actors.

Governments have a special duty to ensure that some portion of the gains of digital enterprise are shared among the public because these enterprises are built upon long-standing collective investments in platforms like the commercial internet, scientific advancement, public education and public roads.

Digital citizenship includes a duty for **organizations, corporations** and **governments** to share the value of the learnings and insights they take from data – large amounts of data or small; anonymized, aggregated or not – with the people who are the sources of that data.

*

The guide for planers and builders of digital systems is modeled, in its form, on the elegant Manifesto for Agile Software Development, put forth by a small group of software developers in 2001 and at its core just four lines long. It advises simple trade-offs in order to get better and more human software made and has since spawned a billion-dollar Agile industry. The guide we offer here aspires to offer a similar path of favoring certain kinds of actions over others, leading to a more effective ethics layer in our digital systems.

This third document aims to offer practical guidance to the architects and builders of digital systems. These are suggestions for making choices that arise in the moment-to-moment technical work of building systems.

GUIDE FOR PLANNERS AND BUILDERS OF DIGITAL SYSTEMS

These positive biases should inform the practical trade-offs that shape the building of digital systems.

A bias toward building and honoring trust with users and
data sources.

A bias toward informed, transparent and consensual use
of data over hidden use of data.

A bias toward sharing learnings and other gains with
data-source individuals.

A bias toward re-training models and expiration dates
over static or perpetual models.

A bias toward minimizing harm over maximizing
effectiveness and efficiency.

- Sadly, this section 2 is deeply uninspiring. The authors raised some interesting dilemmas in section 1 which this rather poor list really fails to answer. Bit of a let down really.

- Suggest Mapping UDHR, ICECR or ICCP rights onto a digital framework

- Really Really Disappointed.

CPSIA information can be obtained
at www.ICGtesting.com
Printed in the USA
BVHW071924090619
550551BV00001B/121/P

9 780578 497631